ALOES
IN SOUTHERN AFRICA

Gideon F. Smith
Braam van Wyk

Struik Nature
(an imprint of Random House Struik (Pty) Ltd)
Reg. No. 1966/003153/07
The Estuaries No 4, Oxbow Crescent,
Century Avenue, Century City, 7441
PO Box 1144, Cape Town, 8000 South Africa

Visit **www.randomstruik.co.za** and join the Struik Nature Club
for updates, news, events and special offers.

First published in 2008
10 9 8 7 6 5 4 3

Copyright © in published edition, 2008: Random House Struik
   (Pty) Ltd
Copyright © in text, 2008: Gideon F. Smith and Braam van Wyk
Copyright © in photographs, 2008: G.F. Smith except for pages
   126, 128, 129, B. van Wyk; page 44, J.P. Roux.
Copyright © in illustrations, 2008: G.F. Smith
Copyright © in maps, 2008: page 9, G.F. Smith and R.R. Klopper;
   page 55, B. van Wyk and G.F. Smith

Publishing manager: Pippa Parker
Managing editor: Helen de Villiers
Designer: Janice Evans
Editor: Gill Gordon
Illustrator: Jaco van Wyk
Cartographer: Martin Endemann
Reproduction by Hirt & Carter Cape (Pty) Ltd
Printed and bound by Craft Print International Ltd

All rights reserved. No part of this publication may be reproduced,
stored in a retrieval system, or transmitted, in any form or by any
means, electronic, mechanical, photocopying, recording or otherwise,
without prior written permission of the copyright owner(s).

ISBN 978 1 77007 462 0 (PRINT)
ISBN 978 1 92054 444 7 (EPUB)
ISBN 978 1 92054 445 4 (PDF)

Front cover: *Aloe excelsa*
Back cover: *Aloe peglerae* (left); *Aloe pluridens* (right)
Right: *Aloe marlothii* (yellow flowers) and *Aloe arborescens*
(red flowers)
Previous page: *Aloe arborescens* x *Aloe ferox*

CONTENTS

Preface 5

Part One: Aloes and their kin 6
Introducing aloes 7
The family Aloaceae and its genera 15
Understanding aloes 29
Aloes and extreme environments 45

Part Two: Aloes by habitat 54
Desert and semi-desert 57
Fynbos (Cape shrublands) 63
Thicket (valley bushveld) 67
Tropical, subtropical and Afromontane forests 73
Grasslands of the Highveld and central interior 76
Savanna (bushveld) 83
Non-discriminating aloe species 96

Part Three: Gardening with aloes 102
General principles for growing aloes 105
Aloes and gardening styles 111
Growing aloes in containers 115
Growing and propagating aloes 119

Part Four: Uses of aloes 126

References 132
Index 134

PREFACE

In this book we present selected aspects of the intriguing biology of members of the genus *Aloe* in Africa and beyond. Commonly known as aloes, we describe and illustrate how they function at both the individual and environmental levels. We also refer to some of their closest and more distant relatives. The intention is to satisfy the curiosity of both the amateur and professional biologist by telling the stories behind these fascinating species and their kin. We deliberately also refer to other representatives that are included in the family of *Aloe* relatives (rather than the *Aloe* family). But this book is predominantly about aloes; stately, sometimes miniature, succulent plants that vividly define the landscapes in which they occur – perhaps the ultimate group of African flagship plants. We also discuss a number of the most common species likely to be encountered in South Africa.

Once equipped with knowledge about aloes, the urge to grow them successfully becomes all-consuming. One of the main reasons is because gardening with succulents in general, but aloes in particular, allows the gardener to break free from the chore of seasonally replacing thick drifts of annuals planted in a neatly raked bed. The predominantly chilli-red and vibrant orange hues of aloe flowers contrast with the verdant greens of their sword-shaped leaves and the surrounding plants. This combination creates a peaceful, calming unity in the garden. In a rockery, or any garden for that matter, shrubby, tree-like aloes form an impressive backbone.

Aloes respond positively to the vernacular African landscape – its prevalent climate and the earthy, roughshod hardscaping accessories created from local materials, like natural rocks and pieces of wood. These sturdy plants embody the creative allure of the African savanna (bushveld) and karroid landscapes, and the satisfaction that comes from creating a natural, indigenous garden. Part of the book is dedicated to the cultivation and propagation of aloes, and numerous tips are given on how to make the most of these plants in human-made environments. Finally, we look at the many uses of aloes today, from contemporary and traditional medicinal preparations to their use in cosmetics, foodstuffs and in the rural environment.

To the team at Struik who worked with us in creating the book – Pippa Parker, editors Helen de Villiers and Gill Gordon, and Janice Evans who designed the book – we extend our grateful thanks.

Gideon F. Smith &
Braam van Wyk
PRETORIA, 2008

The genus *Aloe* displays a wide diversity of different growth forms. However, for most people *Aloe marlothii*, here towering over most of the surrounding plants, represents the archetypal aloe.

PART ONE
ALOES AND THEIR KIN

A form of *Aloe arborescens* with light green leaves and bright orange inflorescences. Taken in the Pretoria National Botanical Garden of the South African National Biodiversity Institute.

INTRODUCING ALOES

Apart from the massive baobab tree (*Adansonia digitata*), the world's largest succulent plant, aloes are certainly among the most familiar of all the conspicuous tall-growing succulents that inhabit the world's arid, and sometimes not so arid, regions. But in contrast to the baobab, aloes are far more diverse and occur over a much broader range of climates and habitats. Wherever the intrepid traveller through Africa may wander, there are sure to be aloes in the landscape. Not all are tall, robust and tree-like; some are small and dainty, and some may hide among rocks or under bushes. Others, perhaps the majority, are medium-sized and can be more accurately described as bulky, thigh-high plants with thick, juicy, persistent and durable leaves congested, often in rosettes, at the tips of branched or unbranched stems.

'Aloe' and 'aalwyn' are the names commonly used for members of the genus *Aloe* in the family Aloaceae. The Afrikaans name *aalwyn*, often used with the prefix *klein-* (small), is applied to aloe relatives, most of which have the appearance of aloes. This family embraces not only the genus *Aloe*, but a number of other genera too, which are discussed in more detail on pages 15–21. As succulent plants, aloes are able to absorb water and store it in the fleshy tissues of their leaves or roots, drawing on it to sustain themselves through dry periods. Aloes are perennial, or polycarpic plants, capable of living for many years and, once mature, of flowering annually.

Stem succulents often cohabit in the same habitats where *Aloe* species and other succulents occur. *Adansonia digitata*, the baobab, has large, chunky but soft stems that resemble lumps of molten candle wax.

ALOES AND THEIR KIN

Today, much of the African interior experiences low precipitation and consequent aridity. Aloes, and other plants found here, have had to adopt strategies that enable them to survive the environmentally hostile conditions. The remarkable diversity of plants found in these parts suggests that large numbers of them have adapted perfectly and not only survive, but flourish in the harsh conditions. The arid parts of southern Africa are host to the largest and richest diversity of succulent plants in the world: an astounding 4 674 plant species, classified in 58 families, show some form of succulence. This means that nearly 47 per cent of the world's succulents grow in southern Africa.

To understand and appreciate aloes we need to become acquainted with as many of their botanical characters as possible, as well as with the animals that feed and breed on them and pollinate their flowers. Varying from compact, miniature rosettes through to majestic trees, aloes demand admiration and respect, and invite closer inspection, touching and smelling. While aloes and their kin form the focus of this book, reference is included to a group of distantly related aloe look-alikes, notably the American agaves. Later sections of the book cover descriptions of selected *Aloe* species, and the cultivation of aloes in gardens.

Succulent diversity:
1. From central Portugal, *Sedum sediforme* has densely arranged, rosulately disposed leaves that are thick with accumulated water.
2. Unrelated species, such as *Crassula rupestris*, also have succulent leaves. The fat, triangular leaves are red-tinged where they are exposed to the African sun.
3. The Euphorbiaceae is a very large family with several thousand species, such as this stem succulent, *Euphorbia mauritanica*.
4. The genus *Aloe* is marvelously variable, but for many, *Aloe marlothii* represents the archetypal aloe.

Where do aloes occur?

Aloes are a diverse group of plants and have spread into almost every possible habitat, from desert and grassland to savanna, and even into comparatively high-rainfall coastal forest. They occur over much of sub-Saharan Africa, ranging from the southern tip of the continent, west and northeast to the Arabian Peninsula. They are also found on the islands of Madagascar and Socotra. With the recent inclusion in *Aloe* of the genus *Lomatophyllum*, its berry-fruited relative, the expanded definition of *Aloe* means that it now also occurs on a number of the Mascarene Islands off the east and southeast coast of Africa.

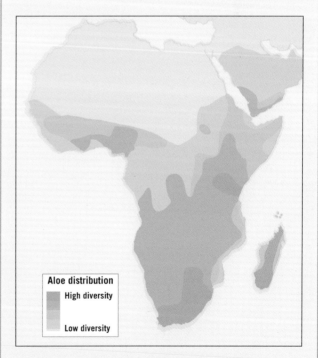

Aloe distribution
High diversity
Low diversity

Aloes are essentially restricted to the African continent, the Arabian Peninsula, Madagascar and some of the islands off the east coast of Africa.

Quick reference to *Aloe* species and their kin

- swollen, succulent leaves that are boat-shaped in cross-section
- leaves usually arranged in a rosette
- large, usually brightly coloured flowers
- candle-like or cone-shaped inflorescences (flower clusters)
- teeth along the leaf margins (generally less vicious than those of *Agave*)

Dominant visual characters

Nearly all aloes have the following visible features:

Leaves that are:
- fat, succulent and boat-shaped in cross-section,
- more or less lance- or sword-shaped,
- arranged into rosettes,
- armed along their margins with sharp or harmless teeth (prickles),
- usually without distinct bud impressions of the marginal spines of adjoining leaves on their surfaces as they unfurl from the centre of a rosette (compare agaves, p 22).

Flowers that are:
- tubular, like jellybeans,
- brightly coloured, but never blue or black,
- borne in densely or sparsely packed, candle-like inflorescences that can be simple or branched,
- mostly borne in winter.

Some common visual characters of aloes include:
1. Boat-shaped leaves (*Aloe globuligemma*)
2. Leaf margins, and often upper and lower leaf surfaces as well, are adorned with teeth (*Aloe marlothii*)
3. Tubular flowers (*Aloe porphyrostachys*)
4. Leaves with white spots (*Aloe squarrosa*)

Three additional characters of aloes and their relatives:

- Most members of the aloe family have a basic chromosome count of eight long and six short chromosomes. The image of the 14 chromosomes (below) contained in a root-tip cell of *Astroloba rubriflora*, a close relative of the true aloes, was taken with a light microscope. The length of the longest of its chromosomes is only about 12 μm (micrometres).

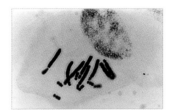

Astroloba rubriflora chromosomes

- Seeds that are, almost without variation, light brown to dark brown or black. This contrasts with Australian *Aloe* look-alike, *Doryanthes*, whose seeds are very light brown.

Aloe suzannae seeds

- Aloin cells in the leaves. Aloin belongs to a group of chemical compounds called anthraquinones, and is one of the main medicinally active ingredients of the bitter yellow juice harvested from the leaves of some *Aloe* species (see illustration top right).

> **NOTE:** To some degree, aloes can be defined by what they are not! They hardly ever have bulbs and do not often form plantlets (bulbils) on their inflorescences. There are exceptions, however: *Aloe bulbicaulis* has a large underground bulb, while *A. bulbillifera* and many other aloes previously placed in the genus *Lomatophyllum* do form plantlets on their inflorescences.

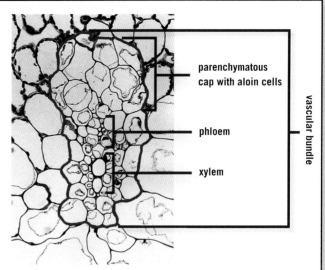

Close-up of a vascular bundle of *Aloe bowiea* showing the cap of parenchymatous cells, some of which are filled with yellow sap containing aloin.

Leaf cross-section

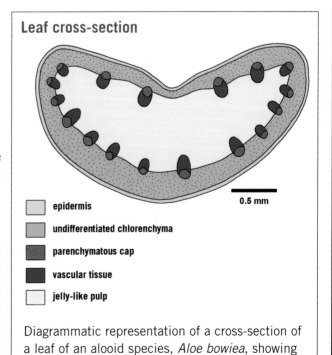

- epidermis
- undifferentiated chlorenchyma
- parenchymatous cap
- vascular tissue
- jelly-like pulp

Diagrammatic representation of a cross-section of a leaf of an alooid species, *Aloe bowiea*, showing the distribution of tissues.

Life span of aloes

In the mid-1970s an investigation into the age of a dead specimen of the desert-dwelling *Aloe dichotoma*, the quiver tree, indicated that the plant could be up to 145 years old (the specimen was estimated to be ± 100 years old, but the methodology used had an uncertainty of ± 45 years). Quiver trees are slow-growing and the largest specimens could be several hundred years old. The greatest ages for individual aloe plants are probably encountered among those that are tall and tree-like, with semi-woody stems. Exceptionally tall specimens (up to 10 m) of *Aloe marlothii* occasionally occur; it is estimated that some of these may be over 200 years old.

Some clonal, non-succulent, desert plants have achieved far greater ages than those recorded for aloes. Following establishment from seed, clonal plants continually produce horizontal stems that give rise to vegetatively produced individuals, called ramets, located at given distances from the parent plant. While individual ramets may be short-lived, a particular clone can live almost indefinitely. An example is *Larrea tridentata*, the creosote bush, the most common shrub in the Sonoran Desert in the USA; some of the plants around today are part of clones that could be over 11 000 years old!

Larrea tridentata, the creosote bush, is named because of its distinctive, lingering scent, particularly after rain.

Useful botanical terms

- **Annual** – a plant that completes its life cycle from germination to seed-set and death within 12 months. Reproduction is the event that ends the growth period, but extrinsic causes, such as frost or drought, may also be involved. Annuals usually have soft, herbaceous stems.
- **Biennial** – a plant that lives for two growing seasons. In the first calendar year it grows vegetatively and in the second it flowers, sets seed and dies. In biennials, flowering is often triggered by the long nights and low temperatures of winter at the end of the first growing season.
- **Perennial** – a plant that lives for three or more years. Because stems in perennials have more time to develop, they tend to become at least slightly woody.
- **Monocarpic** – a plant that flowers and fruits only once and then dies. The term may be applied to annuals, biennials or perennials.
- **Polycarpic** – a plant that can flower more than once (also called iteroparous plants).

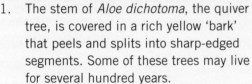

1. The stem of *Aloe dichotoma*, the quiver tree, is covered in a rich yellow 'bark' that peels and splits into sharp-edged segments. Some of these trees may live for several hundred years.
2. The 'bark' around the stem of *Aloe barberae* is grey and lacks the sharp-edged flakes found in *A. dichotoma*. A tangled mass of fibrous conducting tissue is found under the outer layer. *A. barberae* can live for as long as *A. dichotoma*.
3. *Aloe barberae* nears the end of its life span once the rosettes on the tips have become moribund and shrunk in size. This cultivated specimen, from near Uitenhage in the Eastern Cape, is about 80 years old.

Aloes are known to interbreed with great ease. Wherever different species co-occur and flower at the same time, the chances are there will be hybrids nearby. However, some hybrids are not horticulturally successful: the artificially produced *Aloe barberae* x *A. dichotoma* improves very little on either of the parent plants.

THE FAMILY ALOACEAE AND ITS GENERA

The demarcation and definition of the genera within the family Aloaceae have been debated since 1753, when Swedish biologist Carl Linnaeus classified these 'lilies' under the *Hexandria Monogynia* in his book *Species Plantarum*. Linnaeus' system divides plants into 24 classes, according to the number and arrangement of their stamens, and then into orders according to the number of pistils.

Class VI Order I: *Hexandria Monogynia* comprises those species (including aloes) with six stamens and a single ovary. Linnaeus divided the orders into genera, and the genera into species. This early classification system is artificial in the sense that members of groups formed on the basis of a few characters only are often not closely related in an evolutionary sense. Modern systems of classification, on the other hand, strive to demarcate groups based on assumed shared common ancestry.

Even at the rank of family, consensus is yet to be reached regarding the best placement of the group of related alooid genera (*Aloe* and relatives), members of which have been variously classified in the broadly defined Liliaceae, the rather heterogeneous Asphodelaceae, the considerably expanded Xanthorrhoeaceae or, as we prefer here, the narrowly defined family Aloaceae.

A. comptonii is a typical cliff-dweller in its natural habitat.

Ecotypes, chemotypes and cultivars

According to the biological species concept, a species is a sexually reproducing population of individuals. Hence, aberrant (but often horticulturally improved) individuals resulting from hybridization between two aloe species in nature do not qualify for species status, especially if such hybrids are sterile. Genetic (or intrinsic) variation among individuals is a common feature of sexually reproducing organisms and an integral property of a species. Genetically distinct local forms within a species, especially those with a geographical bias, are referred to as **ecotypes**.

Intrinsic variation sometimes manifests at the physiological or chemical level only. For example, populations of *Aloe ferox* in the southern Cape are rich in the chemical compound aloin (with purgative properties), whereas populations further north in the Eastern Cape contain less aloin. Morphologically, however, plants from the two regions appear similar. In such instances the different chemical forms are referred to as **chemotypes**.

The differences between ecotypes and chemotypes found in the same species are retained even if the different forms are grown together in a garden. Given enough time, ecotypes may eventually develop into separate species. If differences between ecotypes (more rarely chemotypes) are marked and there is some discontinuity in variation, they may be formally recognized as different subspecies, varieties or forms of a species.

Horticulturallty superior clones of genetically identical variants are regularly encountered in cultivation where they are being maintained deliberately by vegetative rather than sexual reproduction. Such forms are technically '**cultivars**', not species. In some circles, cultivars are often incorrectly referred to as 'varieties'. Over the years, plant breeders have produced numerous aloe cultivars with outstanding horticultural properties.

Taxonomy and classification

From earliest times humans have been classifying objects naturally and almost instinctively. In order to understand the bewildering diversity of plants and animals, modern biologists have developed means of classifying organisms into groups or clusters ('taxa') according to shared features and assumed common descent. Thus, taxonomy is concerned with classification.

Two steps are involved in the practice of taxonomy. The first is the discrimination of the species – most popularly accepted as a reproductively isolated aggregate of populations that can interbreed with one another because they share isolating mechanisms that, in nature, prevent them from interbreeding with other species. The second step is the classification of related species into a group of a higher ranking. Thus, related species are grouped into 'genera' (singular = genus). Similarly, related genera are grouped into families, and families into orders, orders into classes, and so on to create a hierarchy that in plants ends with the all-embracing 'kingdom' of plants (Plantae). Each plant is allocated a two-part species name that comprises the genus name followed by the specific epithet (for example, *Aloe ferox*).

Genera of the Aloaceae

The family Aloaceae can be divided into five related genera – *Aloe*, *Astroloba*, *Chortolirion*, *Gasteria* and *Haworthia*. Each genus is described and illustrated on the following pages, and a key is provided that will enable easy identification of these genera in the field (see pages 26–27). Bear in mind, though, that in attempting to provide a bird's eye view of the genera of the Aloaceae, some generalizations are unavoidable.

The natural hybrid between *Aloe arborescens* and *Aloe ferox* can grow as a scraggly shrub or as a small tree of over a metre tall. However, its single or multi-branched inflorescences are always fairly long, and the flowers are mostly densely collected into bright orange or red inverted cone-shaped clusters.

ALOES AND THEIR KIN 17

Aloe

The derivation of the word 'aloe' is uncertain, but the following has been suggested:
1. From the Arabic *alloch* or *alloeh*, referring to species used medicinally; a vernacular name for such members of the genus.
2. From the Greek *aloë*, the dried juice of aloe leaves, akin to or derived from earlier Semitic (*alloeh*), Hebrew (*ahalim* or *allal*, i.e. bitter) and Sanskrit words.

The genus *Aloe* comprises more than 550 species of fat-leaved plants that range from trees over 20 m tall to miniatures reaching only a few centimetres above the ground. Most aloes have their succulent leaves arranged into rosettes that are borne at the tips of thin or robust stems. The small, tube-shaped flowers are usually brightly coloured and are carried, either loosely or densely, on single or branched candles that extend well beyond the leaf rosette. The dry leaves of aloe plants are usually persistent on the stems.

Recently, the genus *Lomatophyllum*, originally with about 20 species from Madagascar and the Mascarene Islands, was placed in *Aloe* as a specialized group. This treatment is now widely accepted. In its days of acceptance as a genus, the argument would have been that the berry-like (but still late-dehiscent like a capsule) fruit distinguishes *Lomatophyllum* from *Aloe*, which has capsular fruits that rapidly become woody when they dry out. In general, the leaves of the species formerly included in *Lomatophyllum* are not as succulent as those traditionally included in *Aloe*.

Astroloba

Derivation: From the Greek *aster; astros*, meaning star; and Greek *lobos*, meaning lobe, for the stellately spreading perianth lobes.

The species included in *Astroloba* all have fairly short and thin, but distinct, stems that are borne erectly or sprawl along the ground with age. The leaves of astrolobas are invariably small, triangular and closely packed on the stems. Because of the small size of the sharp-tipped leaves, the plants are widely regarded as a hazard for grazing livestock as the leaves easily get stuck in the animals' throat. The main difference between representatives of the genera *Astroloba* and *Haworthia* lies in the shape of their flowers. The flowers of astrolobas are not as distinctly two-lipped as those of haworthias. Species of *Astroloba* favour, and are restricted to, the arid interior of the southern karroid parts of South Africa.

Like *Lomatophyllum*, *Poellnitzia* is another genus that is no longer recognized by most botanists. It was included in *Astroloba* a few years ago, after making its presence felt in virtually all the other *Aloe*-like genera. It was once treated as a monotypic genus, with *Poellnitzia*

The open flowers of *A. speciosa* are bright white, which contrasts sharply with the red colour of the buds.

rubriflora its only species. Its leaves are triangular and closely resemble those of species of *Astroloba*, but when it flowers there can be no doubt as to why it was regarded as a separate genus for many decades after its discovery. The flowers are pencil-shaped, bright red, and clearly adapted to bird pollination. *Astroloba rubriflora*, as it is now known, is strictly confined (endemic) to the Worcester-Robertson Karoo in the southwestern Cape.

Chortolirion

Derivation: From the Greek *chortos*, feeding place, and *leirion*, lily, for its preferred habitat in grassland.

This genus consists of a single species, in other words it is monotypic. The species, *Chortolirion angolense*, varies greatly throughout a very large geographical distribution range that crosses the southern African winter- and summer-rainfall regions, something that very few *Aloe* species do. In addition, it occurs in desert areas along the west coast, and in tropical habitats in Zimbabwe. It is one of the few bulbous alooids that have a small whorl of deciduous leaves above ground. Its flowers look uncannily like those of some species of *Haworthia*. Two distinct flowering times have been observed for this species: some specimens flower in early spring, usually before the first summer rains have arrived, while others flower in autumn.

Top: The leaves of *Astroloba spiralis* are often grazed off near ground level by domestic stock and game.
Above: In some places, *Astroloba foliolosa* leaves are red.
Right and far right: *Chortolirion angolense* has an onion-like underground bulb; it survives the cold, dry winters of its grassland habitat by shedding its leaves. In the summer-rainfall season, new grass-like leaves (circled) grow above the ground.

Gasteria

Derivation: From the Greek *gaster* meaning belly or stomach, alluding to the swollen base of the floral tube.

Species of *Gasteria* are characterized by their curved, mostly tricoloured flowers with pronounced bulbous swellings. The leaves of most species are picturesquely spotted, with distinct light green or whitish flecks and dots, and they are usually armed with sharp, bony edges rather than with marginal teeth.

It is possible to construct a strong infrageneric classification for *Gasteria* based on the morphology of the flowers, as this group of species is very distinct, especially when in flower, and cannot be confused with any other alooid genera. Indeed, it is probably the only genus in the group for which a strong case can be made for it not to be split or combined with other genera. In habitat, species of *Gasteria* are mostly confined to the shade of sparsely or even densely branched nurse

Gasteria acinacifolia leaves are typically densely mottled. These plants usually grow in the shade of nurse plants.

plants. In nature, very few species venture into fully exposed growing positions. *Gasteria* species favour mild coastal habitats but also occur in some adjacent, climatically less severe inland areas, such as parts of KwaZulu-Natal and the Eastern Cape hinterland.

The flowers of *Gasteria* species have a basal swelling ranging from inconspicuous to pronounced. The flowers of summer-flowering *Gasteria acinacifolia*, a coastal species from the Eastern Cape, are shown here.

Haworthia

Derivation: Named for Adrian Hardy Haworth (1768–1833), English botanist, entomologist, gardener and writer on succulent plants.

Species of the genus *Haworthia* can be identified primarily by means of their rather drab, bilabiate (two-lipped) flowers and small rosettes of highly succulent leaves. The species are mostly small in stature; in fact, with the exception of *H. maxima* and a few species with snake-like, leafy stems, such as *H. coarctata* and *H. viscosa*, most species are less than 100 mm tall when not in flower.

To date three subgenera have been recognized in the genus. The type subgenus, *Haworthia*, consists mostly of the so-called soft-leaved species, many of which have a decidedly blue-green leaf colour. Species of this subgenus have the largest, most beautiful flowers of the three subgenera and they are the only ones whose flowers tend to have distinct tinges of yellow or pink.

In contrast, those species normally included in the subgenus *Hexangulares* are rather stiff-leaved and have less significant flowers. Species of this subgenus tend to be larger than those included in *Haworthia* subg. *Haworthia*.

The third group, *Haworthia* subg. *Robustipedunculares*, includes the largest of the species of *Haworthia*. With the exception of *H. marginata*, the leaves of all the species are adorned, to a lesser or greater extent, with white or concolorous tubercles. The flowers of this group of rather robust plants are the smallest of all the species of *Haworthia*, but are mostly borne on multibranched inflorescences.

Many species of *Haworthia* subg. *Haworthia* (and even *H. bruynsii* of the subgenus *Hexangulares*) have windowed leaves of which the tips are sometimes apically flattened or bent backwards. These species typically grow flush with the ground.

Haworthia fasciata is called 'small aloe' because its growth form resembles that of its larger kin in *Aloe*.

Haworthia glabrata flowers are small and dull-coloured. The free tips are bent backwards like smiling lips.

The aloe-like plants of the New World

Southern Africa has become home to several groups of exotic aloe look-alike plants, most of which belong to the family Agavaceae, which is native to the Americas and some Caribbean islands. Horticulturally, the most popular of these leaf succulents are the frost-hardy yuccas and the massive agaves, or century plants, which live for 15–20 years or more before flowering once and then dying.

Agave

Derivation: From the Greek *agavos* for noble, referring to their imposing stature, especially of the large-growing species when they flower.

The best known of the Agavaceae is undoubtedly *Agave*, a genus of about 250 species that is centred in Mexico. In contrast to *Aloe* species, most members of *Agave* are stemless, although they can grow to enormous proportions. A primary difference between *Aloe* and *Agave* is that species of the latter are exclusively monocarpic – that is, they flower only once before dying. The formation of plantlets on the inflorescences of agaves is a common occurrence, but is rather rare in representatives of *Aloe*.

Furcraea

Derivation: Named for Antoine F. de Fourcroy (1755–1809), a French politician and chemist.

Furcraea is another New World genus that includes several aloe-like plants. Like the agaves, it occurs mainly in Mexico and northern South America. It contains a number of species that can attain tree-like dimensions, but several remain essentially trunkless. Like agaves, they flower only once before dying. In contrast to the agaves, which have erect flowers, the bell-shaped flowers of *Furcraea* hang down from drooping inflorescence branches. In addition, the plantlets produced on the inflorescences tend to be globular, like marbles, rather than distinctly recognizable as immature plantlets ready for the planting.

Above: *Furcraea* species, such as *Furcraea foetida* var. *mediopicta*, look very much like agaves, but their flowers hang down and the plantlets formed on the inflorescences tend to be more globular than those of agaves.
Opposite left: Agaves, or century plants, are monocarpic – that is, they die after having flowered. Some species are perennial through side shoots, such as this specimen of the miniature *Agave nizandensis*.
Opposite right: The large, dull green, clustered flowers of *Agave nizandensis* are borne on a tall, thin inflorescence.

Hesperaloe

Derivation: From Greek *hespera* for evening or night combined with 'aloe' (see p. 18), indicating the occurrence of these superficially aloe-like plants in the American west, where the sun sets.

Once you have seen flowers of *Hesperaloe parviflora*, one of half a dozen species today included in the genus, it is easy to understand why this native of Mexico and the southern USA was at first thought to be a true African aloe. The flowers are pinkish red, borne on tall, branched inflorescences and, at first glance, look almost exactly like those of aloes in terms of architecture. However, the leaves of the species are thin, longitudinally strongly incurved, and adorned with beautifully curled marginal leaf fibres, a character that is absent from *Aloe* species.

Yucca

Derivation: *Yuca*, with one 'c', is a name used for edible roots of cassava, and it could have been misapplied to some yucca species, which have edible flowers.

Members of the genus *Yucca*, although less aloe-like than agaves and furcraeas, can also be confused with aloes. However, their leaves are consistently flatter and thinner than those of aloes. The plants are mostly shrubby, while a few, such as the well-known Joshua tree (*Yucca brevifolia*), can attain tree-like dimensions. Yucca flowers are lantern-shaped and they tend to be adapted for pollination at night. The classic example of mutualism, a type of **symbiotic** relationship, is that between the yucca plant and its highly specialized moth pollinator (see box opposite).

Flowers of *Hesperaloe parviflora* are bright red and closely resemble those of some aloe species. *Dasylirion wheeleri* grows in the background.

Yucca desmetiana leaves are leathery rather than succulent. Interestingly, this species of *Yucca* has never been seen to flower. It grows well in both pot and open-ground cultivation.

Symbiosis

- When species share the same habitat, competition for resources and space can be fierce, but it is not uncommon to find varying degrees of collaboration between different species. Symbiosis refers to an intimate relationship between two or more organisms of different species. Depending on whether the association is advantageous to one or both partners, three broad types of symbiotic relationship are distinguished:
- **Mutualism**: a relationship in which both partners benefit from the association. An example is the close relationship between flowers and specialized pollinators; in the case of aloes, mainly honey bees and/or sunbirds. A particularly close form of mutualism, in which one partner can no longer exist without the other, is the relationship between the yucca plant (right) and the yucca moth.
- **Parasitism**: a relationship in which one member (the parasite) benefits and the other (the host) is adversely affected. Many harmful organisms, including species of insects, mites and fungi, are parasites on aloes.
- **Commensalism**: a relationship in which one organism benefits and the other one is neither harmed nor helped. Examples include epiphytic orchids in the canopy of tree aloes, lichens on the bark of old aloe stems, and birds, lizards, spiders and insects that nest among the dry, spiny and protective leaf remains on the stems of some aloes. In all these examples, the aloe plant is unaffected, while the associated partner derives benefit.

Although they are sharp-tipped, the leaves of the tall-growing Mexican *Yucca elephantipes* are harmless. The inflorescence, with its lantern-shaped, uniformly creamy white flowers, hardly extends beyond the erect young leaves at the tip of the leafy stem.

Key to genera in the Aloe family

1a. Flowers white, whitish or brownish; perianth segment tips distinctly flared open like a gaping mouth, often appearing somewhat irregular or 2-lipped. ➡ **2**

2a. Plants with a typical underground bulb. Leaves hardly succulent, grass-like, deciduous. — *Chortolirion*

2b. Plants without a bulb. Leaves distinctly succulent, not grass-like, persistent. — *Haworthia*

1b. Flowers usually shades of yellow, orange, pink or red, rarely white or green; perianth segments slightly or not at all flared at tips, more or less regular, not appearing distinctly 2-lipped. ➡ **3**

3a. Leaves lacking distinct spines and tubercles; margins white, bone-like, often armed with small semi-translucent, white tubercles. Inflorescences with flowers laxly arranged and usually on one side of the axis only. Flowers hanging down; perianth tube ± curved upwards, lower portion distinctly globosely swollen for up to two-thirds of the perianth length, swollen portion pale to dark pink, sometimes white. — *Gasteria*

3b. Leaves usually armed with spines or tubercles; margins green or brown, not bone-like; if without spines, then leaves usually lance-shaped with faint or prominent longitudinal lines (veins) (e.g. *Aloe striata*), or ± triangular (*Astroloba*). Inflorescences with flowers dense or lax, evenly arranged around axis or, if on one side of the axis only, then usually densely packed. Flowers variously arranged when open, ± straight or slightly curved downwards, lower portion of flower lacking a globose swelling or, if with a swelling (such as in the maculate aloes), then the swollen portion less than one quarter the length of the perianth and the three inner perianth segments fully fused to the three outer segments; perianth in shades of yellow, orange, pink or red, rarely white or green. ➡ **4**

Aloe commixta

Aloe pluridens

26 ALOES AND THEIR KIN

Using a key to identify plants

An **identification key** is a device used to easily and quickly identify unknown plants or animals. One commonly used key – a dichotomous key – presents a series of paired statements that are sequentially numbered. Ideally, only one of the statements in a couplet must be applicable to the specimen that is being identified, while the other of course should not, the former then leading the reader to a further numbered couplet in the key. If a particular statement applicable to the unknown plant leads to a name, then the identification has been completed. A dichotomous key may be likened to travelling a well-marked road that forks repeatedly, each fork bearing directions. If the correct directions are taken, the traveller will arrive at his destination. However, it takes just one incorrect choice for the traveller to become lost, arriving at the wrong destination.

Most keys try to use characters that are readily accessible and easy to observe, either with the naked eye, or with a 10x hand lens or magnifying glass. Be sure to understand the meaning of the terms in the paired statements; do not guess. Always read both statements carefully – even if the first one seems to apply to your plant, the second one may be even better. Constructing an easy-to-use key is not a straightforward exercise; variation within some plant characters tends to be considerable, with characters and their states sometimes difficult to pin down with the required precision. Therefore, do not base a conclusion on a single observation, but arrive at an 'average' by studying several parts of a specimen.

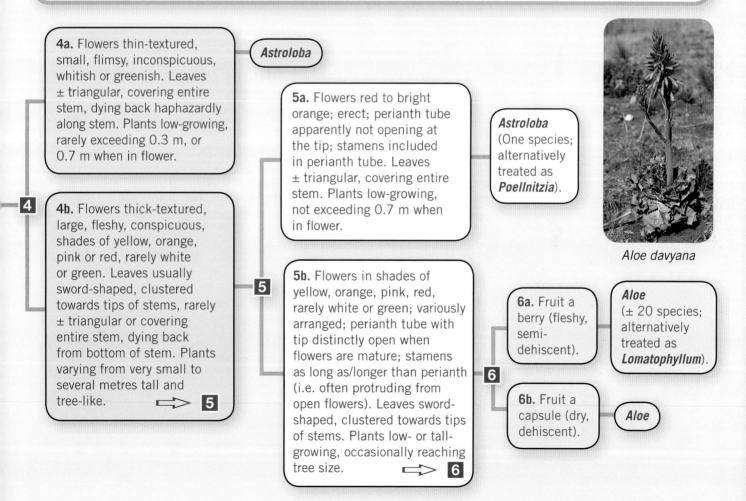

Aloe davyana

4a. Flowers thin-textured, small, flimsy, inconspicuous, whitish or greenish. Leaves ± triangular, covering entire stem, dying back haphazardly along stem. Plants low-growing, rarely exceeding 0.3 m, or 0.7 m when in flower. → *Astroloba*

4b. Flowers thick-textured, large, fleshy, conspicuous, shades of yellow, orange, pink or red, rarely white or green. Leaves usually sword-shaped, clustered towards tips of stems, rarely ± triangular or covering entire stem, dying back from bottom of stem. Plants varying from very small to several metres tall and tree-like. → 5

5a. Flowers red to bright orange; erect; perianth tube apparently not opening at the tip; stamens included in perianth tube. Leaves ± triangular, covering entire stem. Plants low-growing, not exceeding 0.7 m when in flower. → *Astroloba* (One species; alternatively treated as *Poellnitzia*).

5b. Flowers in shades of yellow, orange, pink, red, rarely white or green; variously arranged; perianth tube with tip distinctly open when flowers are mature; stamens as long as/longer than perianth (i.e. often protruding from open flowers). Leaves sword-shaped, clustered towards tips of stems. Plants low- or tall-growing, occasionally reaching tree size. → 6

6a. Fruit a berry (fleshy, semi-dehiscent). → *Aloe* (± 20 species; alternatively treated as *Lomatophyllum*).

6b. Fruit a capsule (dry, dehiscent). → *Aloe*

The leaves of some aloes remain copiously white-spotted throughout. *Aloe squarrosa*, from Socotra, is a good example.

UNDERSTANDING ALOES

Being familiar with and understanding the **parts** that make up plants in the aloe family will help in identification and also cultivation of the species, and will foster an appreciation of their intricate biology.

Roots and anchorage

Aloes and their relatives have an adventitious root system – a network of roots that grow only a few centimetres below the soil surface. This allows the plants to benefit from even relatively low amounts of rainfall. Aloe roots are comparatively soft, which makes them ideal for planting next to man-made structures such as walls and swimming pools. Aloe roots usually do not increase in girth with age and are therefore not strong enough to break concrete. Broken or damaged aloe roots are a bright yellow colour.

Leaves

Although an amazing variety of leaf shapes are found in plants of *Aloe* and its generic relatives, a basic underlying architecture is readily discernible among nearly all the species. Aloe leaves are **succulent** (filled with juice), **boat-shaped** in cross-section and arranged in **terminal clusters**, called rosettes.

Water that accumulates in succulent leaves enables the plants to survive during the dry season. Although the degree of succulence differs vastly among alooid species, most are significantly succulent.

The thin, rather insignificant looking leaves of this grass aloe, *Aloe micracantha*, belie the fact that it carries thick roots underground. This species occurs in the Eastern Cape in grassy fynbos, which is prone to fire and burns easily.

The leaves of most aloes are not fibrous and can easily be broken by hand or cut from a rosette using a knife. In contrast, the leaves of agaves are extremely tough and fibrous, cannot easily be broken off by hand and require a very sharp knife to sever from a rosette.

Aloe leaves are designed to dry out on the plants, often forming a skirt of dried leaves that remains attached to the stems or basal rosettes. In very few cases, as in the tree aloes (for example, *A. barberae* and *A. dichotoma*), old leaves at the base of rosettes are regularly shed as new ones develop, resulting in neat, clean-looking trunks, branches and rosettes.

Leaf shape and arrangement

Leaves of species in the aloe family are mostly boat-shaped in cross-section. By contrast, the leaves of *Kniphofia* species (red-hot pokers) – often mistaken for aloes – are V-shaped in cross-section. Kniphofias have many characters in common with aloes but, with the exception of one species, *Kniphofia typhoides*, they do not have succulent leaves.

One of the most obvious benefits of having boat-shaped leaves arranged in a rosette is that the leaves act as a very effective funnel through which water and moisture are collected and channelled to the plant's roots.

The fan-shaped (distichous) arrangement of leaves, as seen in *Aloe plicatilis* and *A. haemanthifolia*, is commonly found only in the seedling stage of aloes and most of their relatives. But the vast majority of these species lose this character as they mature, with the leaves taking on a distinctly rosette-shaped arrangement. Retaining a juvenile character into the mature stage is referred to as neoteny. Neoteny is seen in some representatives of the related genus *Gasteria*, such as *Gasteria disticha* from the Worcester-Robertson Karoo, some forms of *G. maculata* from the succulent thickets of the Eastern Cape, and the strikingly beautiful miniature *G. armstrongii*, which occurs in grassy fynbos vegetation near Jeffrey's Bay on the southeast coast.

Top and above: The leaves of most aloes are boat-shaped in cross-section, as in *Aloe globuligemma* (top); in contrast, leaves of the red-hot pokers (*Kniphofia* species) are distinctly keeled, as in the leaves of *Kniphofia ensifolia* subsp. *ensifolia* (above).

Although the inflorescences of *Kniphofia* species, such as those of *Kniphofia ensifolia* subsp. *ensifolia*, closely resemble those of aloes, their leaves are consistently less succulent.

The best-known example of an *Aloe* species making a transition from a fan-leaved juvenile to a rosette-shaped adult is *A. suprafoliata*. The common name of this species is 'book aloe', referring to its leaves that in young plants are closely adpressed to one another in two rows, the leaves of the plants looking uncannily like the pages of a book opened in the middle. Plants of *A. suprafoliata* retain this character for longer than most aloes and will even flower while the leaves are still arranged in a fan-shape, but eventually they invariably take on a rosulate arrangement. A fan-shaped leaf arrangement is also found in the miniature *Aloe compressa* from Madagascar, but they retain this leaf arrangement into adulthood.

Leaf spines

The margins of aloe leaves are often armed with spines. These vary considerably from small, soft, harmless and hair-like to large, sharp, pungent and vicious protuberances that can cause serious bodily harm. The upper and lower leaf surfaces can also be variously armed, but this is less common. Leaves of *Aloe marlothii*, in particular, are usually densely covered with spines, but completely smooth-leaved forms of this species are found near Malkerns in Swaziland, among other places. The species with the most attractive leaf spines is undoubtedly *A. aculeata*, where the marginal and surface spines are situated on a prominent white protrusion. With few exceptions, such as the ferociously spined *Aloe ferox*, the leaf marginal spines of *Aloe* species are less harmful than those of *Agave* species, which will easily pierce and tear human skin.

Spots and marks

Many aloes and their relatives have spotted leaves. The marks may vary from very large, white to creamy, H-shaped spots, arranged in merging bands transversely on the leaves, as encountered among the maculate aloes, to minute white spots that are densely and irregularly scattered across both, or only the upper, leaf surfaces.

Above: Leaf margins of several *Agave* species, including *Agave xylonacantha*, are adorned with decorative teeth.
Top: The massive, fibrous leaves of century plants (e.g. *Agave americana* subsp. *americana*, left) were produced to last many years. Agaves store reserves in their leaves to enable the production of a massive tree-like flowering pole rising several metres above the ground-level rosette, such as produced by *Agave gigantensis* (right).

Leaf spots are believed to make it more difficult for grazing and browsing animals to see the plants. This well-known camouflage mechanism occurs in many other species, including aloe relatives. The genus in which it is most effective is *Gasteria*. Both *Gasteria maculata* and *G. acinacifolia*, two common species in Eastern Cape thickets, grow in the rather dense, dappled shade of thorn bush and other shrubs, where the plants perfectly blend with their surroundings.

The considerably smaller leaves of *Haworthia* species are variations of the generally much larger leaves of their bigger counterparts; in all cases, the leaves are more or less triangular or lance-shaped, or thin and grass-like.

It is not only aloes that have leaves that are variously mottled with irregular white spots or sections. This specimen of *Sanseveria hyacinthoides* also has leaves with dense, confluent white bands.

In some species of *Aloe*, the leaves of young plants are spotted with white flecks while those of mature specimens lack spots. Young (top) and mature (above) specimens of *Aloe littoralis* are shown here.

The *Haworthia* species with arguably the most beautiful leaves is the form of *Haworthia maxima* with doughnut-shaped 'pearls' on both leaf surfaces. These bright white structures sometimes have a minute fringe attached to them. Other species, such as *H. attenuata*, *H. glabrata* and *H. radula*, have a multitude of tiny white warts, or tubercles, scattered on both surfaces, which may play a role in camouflaging the plants.

Visual signals, such as those that occur during pollination and seed dispersal, are used by many plants to communicate with and advertise to animals. The white spots or stripes that are associated with species bearing spines on their leaves and stems may perform a similar function in plants, as does the warning coloration found in some dangerous or unpalatable animals. It is possible that once plant-eating animals learn to associate white markings with unpleasant qualities (such as sharp spines or an unpleasant taste), they will avoid those plants displaying them. Conspicuously coloured spines (as occur in many aloes and agaves) may also help to protect the plants by acting as a danger signal to herbivores.

White markings on leaves could also serve as a form of mimicry. The marks may mimic damage done to the leaves by insect herbivores, thus signalling to other insects that the plant has already been under attack. It is known that when herbivores feed on plants, the damaged leaves, as well as adjacent undamaged leaves, become less palatable. Such wound-induced reactions may affect the egg-laying and/or feeding behaviour of leaf-eating insects, causing them to avoid plants, or parts of plants, displaying increased concentrations of deterrent chemical substances. Thus damage mimicry could function as a false-warning signal, provided that the plant is capable of inducing defensive responses to actual leaf damage. Mimicry damage caused by feeding can also function as a potential anti-herbivore adaptation if mimic leaves attract predators and parasitoids that prey on herbivores attacking these plants.

Window-leaves

A window-leaf is one where part of the leaf blade is transparent because the chloroplasts (microscopic organelles that make a leaf appear green) are lacking. Leaf-windows act as effective, if somewhat opaque, filters through which light can reach the inner parts of the leaves where life-supporting photosynthesis takes place. This character is especially useful for plants that grow with their leaves partly sunken into the ground, as it allows a plant to contract into the soil, exposing very little transpiration surface to the atmosphere, yet still carry on with normal, albeit slow, growth. The presence of a window on a leaf may protect its inner tissues against damage caused by the intense light found in arid habitats.

The best known window-leaved species in the aloe family are *Haworthia truncata* and *H. maughanii*; closely related species that occur in the Little Karoo, where they hide in the shade of medium-high shrubs, with only their

Frithia pulchra, a member of the vygie family, has cylindrical leaves with flat, windowed tips.

leaf tips exposed at ground level. These species therefore follow a two-pronged approach to prevent damage to the light-sensitive inner tissues of their leaves.

A number of *Haworthia* species have leaves that are apically flattened and windowed; although a few species of South American cacti grow completely below-ground, none of the *Haworthia* species does so. The occurrence of windows at or near the tips or margins of succulent leaves is not uncommon, and is found among representatives of families as unrelated as Piperaceae, the pepper family (*Peperomia dolabriforme*), the Mesembryanthemaceae or vygie family (*Frithia pulchra* and *Fenestraria aurantiaca*) and Asphodelaceae/Asphodeloideae, the red-hot poker family (*Bulbine haworthioides*).

Direction of leaf growth

The youngest leaves in seedlings and old plants of *Aloe* and their kin are often borne upright, whereas mature leaves tend to grow more horizontally. Having young leaves pointing more or less skywards exposes a smaller surface area of the leaf to direct and scattered sunlight. One consequence is that these leaves receive a lower heat load and so are exposed to less potentially harmful radiation, which in turn promotes leaf development, and therefore plant growth and development (see Seeds and seedling survival, page 43). Once leaves begin to mature, they can more readily tolerate higher heat loads and ultraviolet radiation, resulting in their orientation changing from vertical to more horizontal.

Vertical leaf orientation is also believed to minimize heat loss to the night sky through radiation, thereby protecting sensitive young leaves from nocturnal chilling. Although temperature differences between vertical and horizontal leaves are quite small at night (usually less than one degree), it has been shown in bean seedlings, the leaves of which change orientation from horizontal during the day to vertical at night, that leaf growth can be significantly retarded when the nocturnal orientation of leaves is mechanically constrained to stay horizontal.

Why do aloe leaves sometimes turn red?

The leaves of some deciduous trees change in autumn from green to shades of yellow, orange and red. This happens when chlorophyll, the green pigment in leaves, decomposes to reveal yellow and orange colours (both present in leaves throughout the year but masked by the chlorophyll). At about the same time, various chemical

The leaves of *Aloe dorotheae*, from Tanzania, turn bright red under conditions of environmental stress.

Kalanchoe sexangularis var. *sexangularis'* leaves are known to turn red under environmental stress.

Even if it is grown in cultivation, *Aloe vanbalenii* leaves can turn bright red if the plant is subjected to stress.

processes in the leaves produce red pigments (known as anthocyanins) that accumulate in the cell sap and cause the leaves to turn varying shades of red. The pinks, reds and blues of petals in most plants (excluding, for example, vygies/mesembs and cacti), the red colours of ripe fruit and the deep red skins of wine grapes are due to the presence of these pigments.

Aloe leaves are not shed in the same way as those of deciduous trees preparing for their winter sleep, and their leaves can turn a deep brown or red at any time of the year, regardless of season. Frequently, the onset of red in aloe leaves coincides with stress being imposed on the plants, such as by drought or cold, re-potting or re-planting. (The leaves of some aloes are virtually always red; *Aloe dorotheae* from Tanzania and *A. vanbalenii* from the northeastern savannas of southern Africa are two examples.)

Many aloes from summer-rainfall areas experience their greatest environmental stress during the cold, dry winter season, when most of them flower. High levels of irradiance also contribute substantially towards aloe leaves changing colour from green to brownish red. Anthocyanins are known to absorb light selectively, for example, ultraviolet and green wavelengths, both of which might be harmful to young, sensitive leaf tissues, so these red pigments act as a protective light shield. This explains why the young growth of many plants, especially trees, is often bright red.

In addition to their role as a light-shield, anthocyanins are powerful antioxidants. During stressful periods, plant tissues are subjected to damage by so-called free radicals and reactive forms of oxygen which can damage cell membranes and DNA, often to the extent of causing the death of cells. It is now believed that one of the principal functions of anthocyanins in plants is to remove these free radicals. As human cells can also be damaged by free radicals, it is not surprising that anthocyanins derived from plants are available in certain health foods and medicines.

Tough leaves: cause or effect?

The leaves of *Agave* species are fibrous, strong and sturdy and last for the plant's lifetime. These 'hard fibres' are actually bundles of water-conducting cells (xylem) surrounded by a sheath of individual fibre cells. Most fibre bundles, such as in the sisal leaf (*Agave sisalana*), are located just below the leaf surface and provide structural support that enables the leaf to remain upright.

Hard fibres are used in the manufacture of cordage products (rope, twine, canvas), but can also be pulped for use in speciality papers, such as tissue paper, filter paper, tea bags, currency and security papers.

It is likely that the monocarpic habit of agaves requires that once carbon (the product of photosynthesis) has been expended on the production of structural material, such as fibres and leaf tissues, the plant should conserve its accumulated energy reserves until they are required for reproduction, sometimes up to a decade or more later. Producing the massive flowering pole or 'mast' (inflorescence) requires a lot of energy, structural material and water to be extracted from the fat leaves, the latter becoming deflated like elongated balloons when flowering starts. It is therefore likely that agave leaves are fibrous and strong as a result of the need to reproduce.

Although abundant fibres are not commonly found in aloe leaves, some unrelated species from widely differing areas have been shown to be fibrous. The best known is the aptly named *Aloe fibrosa*, which comes from the Machakos district in Kenya, while two fynbos aloes originating from the southwestern Cape, *A. plicatilis* and *A. haemanthifolia*, also have prominent fibres in their leaves. The functional significance of fibrous leaves in these aloes is not obvious. Elsewhere in the aloe family, fibres are also present in the leaves of *Astroloba rubriflora*, the rather odd species that is sometimes classified in a genus of its own, namely *Poellnitzia*.

Spines, thorns or prickles?

Many plants form sharp-pointed hardened structures, supposedly as a defence against larger animals. Although the term 'spine' is generally used as a collective term for all sharp structures, there are subtle differences, depending on position, arrangement and morphology. In this book, we use 'spine' or 'tooth' to describe any sharp structure, including thorns and prickles.

- **Spines** are derived from stipules and leaves.
- **Thorns** are derived from stems. They are tough structures with their own vascular supply and are usually difficult to break off by hand. ('Spine' is sometimes used for straight structures and 'thorn' for hooked structures.)
- **Prickles** (also called emergences) occur on stems or leaves, usually not in fixed positions. They do not derive from any organ, but are structures that develop from superficially located tissues and are sometimes easily broken off. Typical prickles are found on roses, aloe leaves and leaves of members of the genus *Agave*. Technically, all spines or thorns on leaves of aloes and their relatives are prickles.

Leaf margins and blades of the miniature *Aloe brevifolia* are armed with sharp teeth.

The leaf margins of *Aloe forbesii*, from the island of Socotra, are adorned with creamy white, pointed, but harmless, teeth.

Inflorescences, flowers and fruits

In the aloe family, inflorescences can take the form of many-flowered spikes (inflorescences bearing stalkless flowers), simple racemes (bearing flowers that have stalks), or branched (compound) racemes, known as panicles.

Aloe plants or rosettes do not die after flowering, unlike the look-alike *Agave* species (see page 22). In most aloes, flowering progresses from the base of the inflorescence upwards. True aloes (*Aloe* species) tend to bear densely flowered, highly coloured inflorescences, while representatives of most of the other genera have laxly flowered inflorescences. During the flowering season, aloe plants can be seen from long distances and so are able to attract a variety of animals that can potentially contribute to successful pollination.

The flowers of all species included in the aloe family are tubular. The tubes are formed by six shortly elongated floral (perianth) segments that are variously fused longitudinally. In some species the segments are almost free to their bases (but the flowers are still more or less tubular), while in others they are fused virtually for their entire lengths. These

Aloe vossii flowers are vertically disposed in both the bud and wilted phases, an indication that animals should not visit them. The open flowers turn downwards, offering birds easy access to the nectar.

Flowers of this form of *Aloe cryptopoda* are a uniform bright red colour. The small flying insect (circled) destroys the exterior of the flower base to gain access to the nectar, but does not assist pollination.

tubes are excellent reservoirs for the copious amounts of nectar that most aloe species produce. Since most of the flowers point downwards when they open, the possibility of rain diluting or, worse, washing out the nectar is negligible. Furthermore, the desiccating and contracting floral tube serves to guide and concentrate nectar from the bottom of the flower to the tip where it often accumulates as a glistening droplet. This usually coincides with the female reproductive organs of the plants becoming receptive and possibly offers an additional attractant for pollinators.

The dominant flower colours among aloes are red, yellow and orange. Inflorescences are usually monochromatic, but species with bi- or even tri-coloured inflorescences are also known. In a few species, flower colour can vary within the species, *Aloe arborescens* being one of the best known examples. With very few exceptions, the flowers of haworthias, astrolobas and the single chortolirion are dull-coloured, varying from white to drab brown, green, pink or yellow. Like the aloes, several gasterias have highly coloured red or orange flowers. However, their inflorescences are not

Flower colour varies considerably among aloes. These flowers of the bulbous *Aloe bulbicaulis*, from Zambia, are a dull mustard yellow. In this species, the flowers, flower stalks and inflorescence branches are all the same rather unimpressive colour, a rare occurrence in *Aloe*.

as densely flowered as those of most aloes, making them less conspicuous in their natural habitats.

Once a flower has been successfully pollinated and fertilized, the fruits rapidly develop on the inflorescence, replacing the flowers. Those flowers that have not been fertilized dry out and are soon shed, indicating a lack of nectar or pollen rewards to visiting animals, particularly birds and bees.

The fruits of aloes (called capsules) become woody and dry when they ripen (those of the small group of species previously classified in the genus *Lomatophyllum* remain somewhat fleshy) and then release the usually black seeds. (There are exceptions to this, notably *Aloe variegata*, the seeds of which have prominent white wings.)

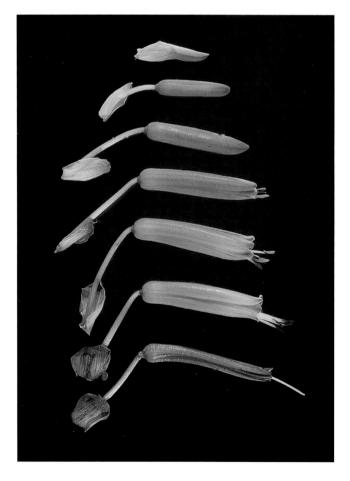

Above: Aloe flowers don't always open from the bottom of the inflorescence upwards. This specimen of *Aloe speciosa* seems unsure of what is expected of it when it comes to flowering sequence.
Left: Flowering sequence of the registered cultivar *Aloe* 'Spiraal'. The flowers are protrandric, meaning that the pollen matures before the stigma becomes receptive. Here, the stigma is exserted beyond the mouth of the flower in the bottom-most flower.

Aloe parvibracteata flowers often have a conspicuous, nectar-filled basal swelling.

Bees frequently visit *Aloe marlothii* flowers, possibly just to collect pollen rather than contribute to pollination.

The spent flowers of *Aloe porphyrostachys* remain closely pressed to the inflorescence axis.

The fruits of *Aloe marlothii* are light brown, tinged with green. The irregular bumps on the fruit surface are a sure sign that the capsules have been parasitised.

Fruits of *Aloe broomii* var. *broomii* are light green and so tightly packed that the dry floral remains are retained between them.

The closely packed orange-brown fruit capsules of *Aloe ferox* are almost as decorative as its flowers.

Seeds and seedling survival

Aloe seeds are almost invariably quite small and black. They are angled, slightly thickened in their centres, and have short or prominent wings which, although quite rudimentary in most species, are believed to facilitate wind dispersal of the seeds.

As seedlings, some species of *Aloe* often start life in the shade of 'nurse', or companion, plants. This is critically important as, during the first few months of their lives, the young plants are very tender with comparatively weak root systems, making them prone to a multitude of setbacks. These include being grazed by game or domestic livestock, or being attacked by parasites such as red spider mites (a type of arachnid) and sap-sucking insects. Most aloes soon outgrow their nurse plants and tower over them, making it difficult to imagine that they initially depended on these companion plants for survival. But some, such as the grass aloes, never outgrow or overtop the plants with which they co-habit. Not only do their leaves resemble those of the grasses in which they grow, the aloe plants are also the same size as their companions. (See also page 77.)

Nurse plants provide the initial shelter required by young, tender seedlings of a variety of species, including succulents. In deserts in particular, succulents depend on nurse plants for survival during the establishment phase. The seeds of many aloes germinate underneath grass tussocks or low-growing shrublets, the latter serving as nurse plants which protect the young seedlings from excessive heat and radiation, desiccation, frost and plant-eating animals.

Seed of *Aloe spicata* germinate profusely under the protection of fallen tree leaves but, ultimately, some of these seedlings will succumb to predators, with only one or two reaching maturity.

Gregariousness among aloes

Individuals of many species of *Aloe* grow closely together in very dense stands, sometimes forming thickets. Examples of species that exhibit such gregariousness include *A. africana*, *A. arborescens*, *A. ferox*, *A. marlothii*, *A. microstigma*, *A. spicata*, and several others. In the case of many other plant species (with the exception of most wind-pollinated species), this is a rather uncommon phenomenon, as plants produce chemical signals that prevent the seeds of the same, or other, species from germinating in close proximity to them. Therefore, in terms of plant dynamics, spaces between plants of a certain species that remain unoccupied can be as informative as the occupied ones, regarding the structure of a plant population and the interaction among individuals in it.

Not a common sight: *Aloe microstigma*, growing here in its natural habitat in Verlatekloof, Northern Cape, had to endure a fall of snow during the winter of 2006.

ALOES AND EXTREME ENVIRONMENTS

Aloes and fire

Every natural habitat in which sufficient combustible material is produced has its own characteristic fire regime. These habitats range from being highly susceptible to fire to essentially fire-free.

In southern Africa many species of *Aloe* occur in habitats that can be regarded as fire-free. This includes parts of the Great Karoo, where the availability of fuel is insufficient to sustain fire because the distance between adjacent plants is too great and the production of material that can burn is too low. *Aloe* species also occur as a component of succulent thicket vegetation, where many plants with fat, moisture-filled leaves make it difficult for fire to take hold. This is often the case in the hot, arid valleys of east-flowing rivers in the Eastern Cape and KwaZulu-Natal.

However, some habitats in which aloes occur *are* prone to regular, almost predictable fires. The Cape Floristic Region, which coincides closely with the Fynbos Biome, is the best known of these, but the grassland and savanna regions are also fire-prone.

Under normal conditions, fynbos vegetation burns at a frequency of between five and 20 years. However, as a result of recent human impacts, such as the introduction and subsequent uncontrolled spread of alien vegetation, the *frequency* and *intensity*

The outer leaves of this specimen of *Aloe peglerae*, growing in a protected pocket between some rocks, were scorched by a fire, but the damage did not prevent the plant from flowering.

of fires have increased dramatically, with disastrous effects on property and the natural environment.

Fynbos vegetation can be divided into various subtypes, depending on altitude and a number of climatic criteria. On a broad level, however, we can distinguish mountain from lowland fynbos. Comparatively few aloes occur naturally in mountain fynbos, notable exceptions being the tree-like *Aloe plicatilis* and the low-growing *A. haemanthifolia*. Both of these, rather uncommonly for species of *Aloe*, have tongue- or strap-shaped leaves arranged in a fan-like manner, but they differ considerably from one another in terms of growth form and height. *A. plicatilis*, commonly known as the fan aloe or *waaier-aalwyn* in Afrikaans, is a sparsely to much-branched tree, carrying its rosettes at the tips of the branches. One adaptation to surviving in a fire-prone habitat is seen in the stems and branches of this species, which are protected by a thick, corky bark, a very uncommon character in the Aloe family. In addition, *A. plicatilis* favours rocky and scree habitats where hot fires generally cannot reach the plants.

Aloe haemanthifolia, on the other hand, is a low-growing, more or less stemless herb. It has an underground rootstock with a strong root system from which new stems and leaves develop (sprout) if the above-ground parts are ravaged by fire. Although the above-ground growth may be completely destroyed by fires, plants are rarely killed. The plant is therefore a resprouter, one of the more effective strategies (a common alternative strategy, though not occurring among aloes, is displayed by so-called reseeders: plants are killed by fire and must regenerate anew from seed) evolved by plants in the fynbos (and elsewhere) to survive regular fires. The shrubby *A. commixta* also has a strongly developed underground rootstock and roots, and is a resprouter. So too is *A. micracantha*, the only species of grass aloe that occurs in grassy fynbos in the Eastern Cape province. It has a rootstock with exceptionally thick and robust roots, so much so that the below-ground biomass (mass of living matter) of this species by far outstrips the above-ground biomass (as you discover when trying to remove a plant from the ground without damaging its spindle-shaped roots!). In this species, new rosettes develop from the rootstock if the plants are burned or, as often happens in urban open spaces, are cut off at ground level with industrial-strength lawn-mowers.

Aloes found in lowland fynbos, including the creeping *Aloe perfoliata* and the miniature *A. brevifolia* from the chalk fields near Bredasdrop, have developed different strategies for survival. They evade fire through a preference for rocky habitats that are often, at best, only sparsely covered with vegetation that can act as fuel during a fire in the dry summer season. This strategy is also found in the summer-rainfall region, where a number of grass aloes, such as *A. verecunda*, are inclined to grow in protected rock crevices.

The dried leaves of the miniature *Aloe petrophila* remain attached to the stem, even in a habitat that is well protected against fires. In the dry winter season, the leaves turn a deep purple, contrasting with their white leaf spots.

Above: *Aloe marlothii's* 'skirt' of dried leaves around its stem may play a role in insulating the stem during fires.
Top: The corky bark that covers *Aloe plicatilis* stems provides protection against the heat of the fires that regularly ravage the southwestern Cape fynbos.

But fynbos is not the only habitat prone to regular fires. Grassland and savanna (bushveld) are too, and here the grass aloes survive fire through strongly developed rootstocks from which new plants grow in the ensuing rainy season. A miniature relative of *Aloe*, *Chortolirion angolense*, which has rather insignificant haworthioid flowers, survives fire by resprouting leaves from a typical underground bulb-like structure. The renewal (perennating) structure of this species is as close to a conventional bulb as one will get in the *Aloe* family.

Among the possible reasons for the retention of a skirt of dry leaves around the stems of single-stemmed, tree-like aloes is that these act as insulators against the heat of fires. *Aloe ferox* and *A. marlothii* are a case in point, as both species grow naturally in areas that carry sufficient fuel to ensure the occasional occurrence of warm to hot fires.

Fire – good or bad?

In spite of having developed various strategies for surviving fire, all aloes, especially the medium-sized ones, can be very detrimentally affected by fire. If a fire that is too hot occurs in the flowering season, the inflorescences will be seared and no fruit will set. Furthermore, many winter-flowering aloes occur in winter-dry savanna and grassland habitats, precisely the time of year when fires are most likely to occur. And of course, in some cases, aloes – and their relatives – may simply be killed outright by fire.

But then there are also some aloes that benefit from fire. Seemingly, some grass aloes will not flower unless they are exposed to fire. The most notable of these is *Aloe chortolirioides* which may not flower at all during years when the grassland remains unburnt. In addition, the subsequent release of nutrients into the soil through ash production enriches the soil, thereby enhancing seed germination, among other things. A grassland fire, followed by good early summer rains, is sure to stimulate extensive flowering in many of the grass aloe species. An indication that grass aloes are dependent on fire is evident in the condition of their dead winter foliage, which burns with ease (as does the grass in which they occur). The ideal timing of fires for the grass aloe species is late July and early August, just before the beginning of the growth cycle.

This undescribed *Aloe* species, growing in Limpopo Province, survived after being exposed to a grass fire, and still flowered shortly thereafter.

Mediterranean climate and winter-rainfall deserts

A Mediterranean climate is one in which the summers are hot and dry and the winters wet and mild, such as is experienced in parts of southern Europe (macchia/maquis vegetation), southern California (chaparral), the Western Cape (fynbos), southern Australia (kwongan) and Chile (matorral). Within these relatively high-rainfall climates, aloes occur naturally only in the southwestern Cape, although many aloes flourish in cultivation in the other areas mentioned. In contrast to other 'Mediterranean-type' regions, the Cape experiences additional precipitation, albeit in limited form, from rain and mist throughout the warm, dry summer months, especially at higher altitudes on the Cape Fold Belt mountains. Furthermore, water released from higher up the surrounding Cape mountains creates a steady run-off, even in the driest months.

The designation 'Mediterranean climate' is not applied to all regions with predominantly winter rainfall. Regions that experience very low winter rainfall are referred to as winter-rainfall deserts. Such arid regions include the Succulent Karoo (incorporating Namaqualand) in southern Africa and the desert of Baja California in North America. Winter-rainfall deserts are very rich in succulent plants, but the Succulent Karoo is the only one that harbours aloes (about 20 species, including the tall and sparingly branched tree aloe, *Aloe pillansii*).

In Lesotho, where winter temperatures regularly drop below freezing, *Aloe striatula* is often planted as a hedge. Here, *Agave americana* subsp. *americana*, a native of Mexico, grows behind the hedge.

Aloes and cold tolerance

In general, succulents are surprisingly cold tolerant. This certainly also applies to species of *Aloe*. Exceptions to this rule are those species whose natural distribution is restricted to subtropical and near-tropical areas. *Aloe thraskii*, which occurs naturally along the humid subtropical coastline of KwaZulu-Natal, is such an example. It invariably requires protection against the mild to severe frosts that are prevalent on the Highveld during the winter months. In contrast, species that occur in high-lying parts of the subcontinent, for example, in the Drakensberg of the Eastern Cape, Lesotho and Mpumalanga, as well as in the Great Karoo, can tolerate very low temperatures. Some species have a distribution that straddles areas that are both frost-free and subjected to either light, mild or severe frost. In such instances the species may well have local forms differentiating into distinct ecotypes (see page 107).

Ice formation in plant cells under freezing conditions

Unable to retreat into a burrow or migrate to warmer climates, plants seem especially vulnerable to cold. Unlike warm-blooded animals, plants are also not able to maintain (thermoregulate) their tissues at a constant temperature on their own (except in a few highly specialized flowers). Yet, many survive freezing temperatures for months on end. Damage to plant tissues by extreme cold is not due to the low temperatures *per se*, but is usually caused by sharp ice crystals that form inside cells with resultant damage to cell membranes. To increase cold-tolerance, cell membranes in some chilling-resistant (or cold-acclimated) plants are chemically different from the membranes in chilling-sensitive plants. Membranes in such cold-adapted plants are able to maintain membrane flexibility to much lower temperatures, thus protecting the cell and its contents from damage by ice crystal formation.

In another strategy, some plants, including members of the aloe family, produce antifreeze compounds that protect cells against intracellular ice formation. Some of these compounds (sugars, amino acids and other solutes) induce supercooling in the plant's tissues by inhibiting the growth of ice crystals. Supercooling refers to water in cells that is maintained in a liquid state below 0°C.

Aloe polyphylla, which can grow to the size of half a wine barrel, is indigenous to Lesotho, one of the coldest areas in southern Africa. It can easily withstand subzero temperatures.

Contrary to what we would expect, the freezing of water may, under certain conditions, help to protect cells from frost-damage. When water changes from a liquid to a solid (ice) during freezing, energy in the form of so-called latent heat is released. As the temperature drops below freezing, living plant cells lose heat, but solutes prevent the watery contents of the cells from forming ice crystals and the liquid supercools. However, as the temperature drops further, ice starts to form in the spaces filled with moist air between the cells. As the ice crystals grow, water (but not any solutes) moves out of the cells through the cell membrane and porous cell wall and adds to the growing ice crystals outside the cell. This slow drying out (dehydration) concentrates solutes within the living cell contents (protoplast), resulting in a lowering of the freezing point further by 2–3°C. Ice formed outside a cell (extracellular ice) does not kill plant cells. As ice forms in the spaces between the cells, heat energy is released and the temperature of the liquid inside the cells rises a few degrees and stays at the higher level until all the extracellular spaces are frozen. When that point is reached, latent heat release stops and the temperature begins to fall again. When the ice outside cells melts in frost-hardy plants, the water goes back into the cells and they resume their normal metabolism. The release of heat energy during ice formation is the basis for the common practice of spraying plants with water during frost. As long as the water continues to freeze outside cells, it releases heat that prevents freezing inside cells.

The flowers of this specimen of *Aloe variegata* were destroyed by frost during a cold snap.

Because of their high water content, most succulents are particularly prone to damage by subzero temperatures. Yet, many aloes are surprisingly cold-tolerant. Worldwide, succulents hardly ever occur naturally in areas experiencing *severe* frost, especially during the season of active growth. On high mountains, succulents tend to decrease in abundance with altitude and are virtually absent above the mean cloud line where severe night frosts regularly occur throughout the year. *Aloe polyphylla* (spiral aloe), the national floral emblem of Lesotho, is unusual among succulents in general and aloes in particular for being adapted to flourish in the extreme cold of its mountain habitat above 2 400 m in Lesotho. Plants are often covered by snow in winter. The specific mechanisms employed by this remarkable aloe to avoid lethal frost damage are not known.

What is frost?

After the sun has set, the land radiates heat that it has stored during daylight hours, and soil and plants eventually cool to a point at which they are colder than the surrounding air. At this point (dew point), any moisture in the air condenses in the form of water droplets (dew) on plants (and, of course, on other structures, such as outdoor garden furniture, vehicles etc).

If the air humidity is very low, i.e. if the air is dry, the dew point will be reached at a much lower temperature, closer to freezing. This typically occurs in inland, continental areas away from the coast. If the air temperature drops below freezing, the moisture in the air is deposited in solid form as frost. Thus, frost is a solid deposition of water vapour from saturated air, and NOT, as many people assume, dew that has subsequently frozen. Since a small amount of heat is released as the water vapour converts to frost, the temperature of the plant sap usually stops dropping for a short while and the visible frost on the plant surface provides some 'cold protection', especially if the air is reasonably moist.

Black frost occurs on very cold nights when the air humidity is so low that 'protective' visible white frost does not actually occur. Under such conditions the temperature of the leaves of plants, including aloes, drops to that of the air and the plants lose moisture to the air through evaporation, which further lowers the temperature of the leaves. This results in the freezing of sap inside the cells of susceptible plants. Moisture loss and evaporation is even greater when a strong, cold wind is blowing at the time. These conditions result in severe freezing damage. As the frozen sap of plants melts when temperatures rise again, damage becomes visible as a blackening of dead plant parts (hence the term 'black frost').

To prevent frost damage to flowers and inflorescences, some forms of species (ecotypes) that are generally not frost resistant are genetically programmed to postpone their flowering events to fall outside the cold winter months. This is particularly evident in summer-flowering forms of *Aloe ferox* from parts of the Northern and Eastern Cape and southern Free State, and spring-flowering forms of *Aloe marlothii* from the Winterton and Bergville areas in KwaZulu-Natal.

Aloe intelligence – do plants respond to environmental conditions?

Sometimes, in the flowering season, you may encounter an aloe population with not a single flower in evidence. There are several potential reasons for this: the flowers may have been removed by baboons, who destructively break off inflorescences in search of nectar; the preceding growing season may not have been favourable for plant growth, resulting in a shortage of water and food reserves to invest in flowering and fruiting; or the plants sensed an oncoming event, such as a drought or severe cold snap, that could destroy any flowers produced. Plants may also 'deliberately' deprive destructive predators of a season's worth of food, in the form of flowers and seeds, so as to minimize the impact of such pests on subsequent seed crops (see 'masting' on page 53).

So, do plants have built-in 'intelligence' that can anticipate environmental conditions that could be conducive to the destruction of the plant and/or its flowers, fruits and seeds?

A well-established phenomenon in many perennial plant populations, especially trees, is a reproductive pattern wherein an entire population of one species reproduces heavily at once, but at long intervals.

Such episodic productions of large seed crops result from mass flowering. Referred to as **'masting'** or 'mast-fruiting' by biologists, such events are generally followed by one or more years of comparatively poor flowering and/or seed set.

It is believed that masting plants put so much energy into producing a good seed crop that they are forced to forgo reproduction in at least one subsequent year. Although masting has never been described in aloes, it is possible that the non-flowering behaviour observed in groups of aloes may be related to comparable periodic cycles of an exceptionally good seed crop in one year, followed by one or more years of relatively poor crops.

One of the most popular theories to explain masting is that in 'mast' years, the bumper crops satiate seed eaters, providing them with so much food that some seeds escape being eaten. Furthermore, the small crops produced in subsequent seasons serve to reduce populations of seed predators, resulting in fewer animals to eat the seeds produced during mast years. Thus, a higher overall proportion of seeds ultimately escapes predation. Another suggestion is that masting might enhance pollination efficiency, especially in wind-pollinated trees. It has also been proposed that the year after a period of water stress (or other suboptimal conditions for growth), forest trees may respond with high seed production. Although the cues that trigger masting are poorly understood, past and present weather conditions, especially fluctuations in temperature, seem, in some instances, to correlate with this still largely mysterious ecological phenomenon. But, what about the possibility that such synchronized reproductive patterns in plants may also be *in anticipation* of future climatic conditions?

Traditional methods of predicting weather often make use of animal behavioural patterns, and in recent years even scientists have been studying animal behaviour in the context of predicting natural disasters. Furthermore, it has long been observed that weather change over the short term is associated in humans with increased complaints, especially of chronic pain. Stormy weather causes the air pressure to drop and then rise as the storm moves past. Experiments based on animal behaviour have shown that low barometric pressure and low ambient temperature augment pain intensity. These observations support reports from humans with certain pathological conditions that pain is aggravated by an approaching low-pressure system or exposure to a mildly cold environment.

If some humans can predict approaching inclement weather in this way, then could plants do the same? Currently, there is no scientific evidence to support this. It is nevertheless widely claimed by gardeners (at least in South Africa) that the so-called rain or storm lilies (*Zephyranthes* spp.) of South America display mass blooming a few days *before* such stormy weather. On the other hand, reports in the popular literature state that such flowering only happens *after* good rains. It would be informative to observe a stand of these lilies and to record weather patterns over one or more seasons to establish the facts and whether they can predict certain weather conditions.

Plant behaviour is still very poorly understood. In folklore, influences of the phases of the Moon on plants have long been claimed. Although still treated as myth by many scientists, some of these assertions may well have some scientific foundation. Using sensitive measuring equipment, scientists have shown that the Moon does influence the flow of water between different parts of a tree. Moreover, plant neurobiology, a newly emerging field of plant sciences, depicts plants as social organisms with surprisingly complex forms of behaviour. In their own special way, plants can see, smell, taste, touch and, perhaps, even hear. They have the power to compute, they show foresight and they remember what happens to them – all qualities associated with intelligence in animals!

PART TWO
ALOES BY HABITAT

For the purposes of this book, the southern African landscape has been divided into six broad regions according to dominant vegetation and climate. A selection of representative aloe species is discussed for each region, providing the reader with an indication of typical habitat and natural growing conditions for the different species.

The six regions are: desert and semi-desert, fynbos (Cape shrublands), thicket, forest (temperate, tropical and subtropical), grassland (temperate, tropical and subtropical) and savanna (bushveld). A seventh category discusses non-discriminating aloe species that grow naturally across a number of regions.

Aloe brevifolia, a dwarf fynbos species.

Aloe perfoliata, a creeping fynbos species.

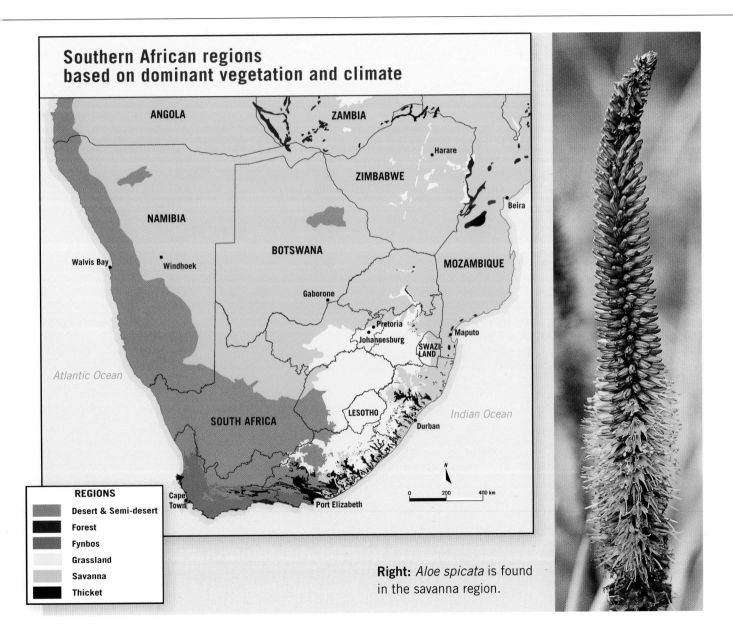

Southern African regions based on dominant vegetation and climate

REGIONS
- Desert & Semi-desert
- Forest
- Fynbos
- Grassland
- Savanna
- Thicket

Right: *Aloe spicata* is found in the savanna region.

ALOES BY HABITAT

The flowers of the desert-dwelling *Aloe dichotoma* are a bright, butter yellow, while the exserted stamens are a contrasting purplish colour.

DESERT AND SEMI-DESERT

The term 'desert' is not widely used to describe the often exceptionally dry western parts of southern Africa. For example, we virtually never hear reference to the 'Richtersveld Desert', just the Richtersveld. Even the Karoo, of which there are several distinct regions, is not called the Karoo Desert, despite the word being derived from the Khoekhoe word '*garo*' meaning 'land of thirst'. In South Africa, semi-desert conditions are often described as 'karroid' conditions. Locally, 'desert' tends to be reserved for arid and hyper-arid sandy regions, like the Kalahari and the Namib, respectively. Note, however, that because of its often tree-dominated vegetation cover, the so-called Kalahari Desert (covering most of southern and central Botswana) is best described as savanna, as is shown in the map on page 55.

In southern Africa, desert-like areas occur in both winter- and summer-rainfall regions, as well as in regions that receive sparse rainfall at any time of the year. The winter-rainfall dry areas (or deserts) are probably the best-known, thanks to the mass displays of spring wildflowers in Namaqualand (see box on Mediterranean climate, page 48). The most common contributors to this floral showcase belong to the Asteraceae, the daisy family, while among the succulents, the Mesembryanthemaceae (vygie family) contributes a significant number of species. However, over 20 per cent of southern Africa's *Aloe* species occur here.

Apart from the well-known Namib and Kalahari deserts, large parts of southern Africa are subject to low rainfall and reasonably predictable droughts. Although this landscape, viewed towards the Witteberg mountain range from Whitehill, a small railway siding in the Karoo, appears to harbour a low diversity of plants, it is a succulent paradise, with numerous miniature species hiding under the low-growing shrubs.

Aloe dichotoma

This is perhaps the best known of all the tree aloes as it often occurs in photographs depicting the Namaqualand or Namibian landscape. It is a single-trunked species with a large canopy that often looks over-sized. The canopy consists of a few to many branches, each of which supports a small rosette of dull green leaves that contrast sharply with the butter-yellow flowers produced in winter. It is not the easiest of the tree aloes to keep alive outside its natural habitat.

Aloe ferox

These stately, mostly single-stemmed aloes are found in the arid karroid interior. The leaves are distinctly boat-shaped and gracefully curve upwards or outwards. Stems are typically clothed in a skirt of dry leaves. Flowers are arranged into long candles, carried in winter. Flower colour varies considerably, from red through orange, pinkish, to yellow and even white. The thick, juicy leaves contain several compounds that are used in the cosmetics and pharmaceutical industries.

Flower colour varies considerably in *Aloe ferox*, particularly in the Eastern Cape, but the orange form is the most common.

Aloe pillansii

A magnificent tree aloe, although not nearly as common in its natural habitat or in cultivation as *Aloe dichotoma*. This is one of the most threatened of the southern African aloes. It has a single stem and a few large, robust branches that support smallish rosettes of leaves at their tips. The inflorescences grow from near the bottom of a rosette, first curving downward before they reach skyward in a rather lazy-looking way. The dull yellow flowers are nearly round, at least in bud stage.

Aloe ramosissima

In contrast to *Aloe dichotoma* and *A. pillansii*, this is usually a robust shrub. Its main stem is normally very short and supports a large, dense canopy consisting of a multitude of small rosettes. The leaves are thick and dull green, but generally somewhat smaller than those of *A. dichotoma*. As is the case with *A. dichotoma*, *A. ramosissima* has bright, butter-yellow flowers.

Above and below: *Aloe ramosissima*

Aloe variegata

Plants do not grow very tall and consist of small, usually solitary, sometimes clump-forming, rosettes. The deep-green, triangular leaves with white markings look much like those of species of *Gasteria* and are densely packed into overlapping layers. Large, pinkish red flowers are carried in laxly flowered inflorescences. This was, and perhaps still is, one of the most popular aloes grown in the temperate climates of Europe and the UK, especially in greenhouses, on suburban windowsills and in the dusty windows of pubs.

Shrublands, such as the fynbos of the western and southern Cape, are not known as an *Aloe* paradise, yet a handful of species are indigenous to the region, including the fire-adapted *Aloe plicatilis*. Pictured here is an area near Kommetjie, in the Cape Peninsula, which is rich in a multitude of shrubs, including several *Erica* (heather) species.

FYNBOS (CAPE SHRUBLANDS)

The Cape Floristic Region (CFR), also known as the Cape Floristic Kingdom or simply 'Capensis', is situated at the southernmost tip of Africa. It stretches in an L-shape from the vicinity of Vanrhynsdorp and Nieuwoudtville in the northwest along the west coast southwards towards Cape Point, a distance of about 250 km, and then in broken strips and patches eastwards to near Grahamstown, a distance of about 800 km. In the extreme east the CFR peters out into and mixes with subtropical thicket, coastal forest, grassland and karroid vegetation.

Covering an area of about 90 000 km^2, the CFR is the smallest of the world's six Floristic Kingdoms, but has the richest and most diverse flora by far, a total of over 9 000 species. Furthermore, most species (more than 6 000) are endemic (strictly confined) to the region. Most of these are associated with a dense type of evergreen shrubland, known locally as fynbos. This vegetation type is very characteristic of the CFR, and the region is sometimes referred to as the Fynbos Biome.

The climate prevalent in the CFR is generally described as Mediterranean (see page 48). This means that the winters are mild and wet and the summers are hot and rather dry, although especially towards the east, thunderstorms are not uncommon during the warmer months of the year. Frost is restricted to the inland valleys, and snowfalls to the higher mountains. Mist moving overland from the Atlantic Ocean frequently occurs over the west coast of South Africa.

The nutrient-poor status of the sandy soils mainly associated with quartzitic sandstone mountains of the CFR is the most outstanding pedological feature of the region. Fynbos is fire-prone; indeed, the regeneration of many species, especially the so-called reseeders, is impeded if fire is excluded for too long.

Apart from three well-known families – Proteaceae (proteas, pincushions, conebushes), Ericaceae (ericas or heaths) and Restionaceae (restios or reeds) – that occur in abundance in the area, the CFR is home to a large number of succulents representing the Mesembryanthemaceae (vygies or mesembs), Geraniaceae (pelargoniums) and Crassulaceae ('plakkies' or stonecrops).

Not many aloes grow in fynbos, but those that do have become very popular in amenity and domestic gardening in those parts of the world that experience a Mediterranean climate.

Aloe brevifolia

This miniature aloe suckers extensively by sprouting rosettes from its base to form large clumps (clones). Some forms of the species can grow quite big, but in horticulture the small, typical form is the most popular. The bluish green leaves are short and distinctly angled in cross-section. The large red flowers are sparsely to fairly densely spaced in a once- or twice-branched inflorescence.

Aloe commixta

This creeping species has comparatively thin stems that are not able to support themselves in an erect position unless they scramble into surrounding vegetation. The leaves are widely spaced on the stem and only slightly succulent. The bright yellow flowers are fairly large, but the flower clusters are quite small.

Aloe comptonii

Plants form fairly long creeping stems that retain the remains of dry leaves, but these whither more quickly than the leaves of most stem-forming species. Leaves are bluish green and carry numerous stubby teeth on their margins. Flowers are arranged into large, head-shaped inflorescences. This species mostly occurs on the margins of fynbos patches in the Little Karoo, where plants often dangle down near-vertical cliff-faces.

Aloe perfoliata

Plants form long stems that creep along the ground. The stems are covered in green or dry, short, more or less triangular, leaves that point upwards. The inflorescences are curved upwards and vary from head-shaped to conical. The species is commonly cultivated on the Mediterranean coast of Europe, where it flourishes in the prevalent climate which is so similar to that of its natural habitat.

Aloe plicatilis

The basal and lower parts of the stems and branches of this tree-like aloe are covered by a strikingly corky bark. The tip of each stem carries a cluster of tongue-shaped bluish green leaves, which are eye-catching as they are arranged in the shape of a fan, forming a distinct row as seen from the growing tip of the shoot, usually with 180° between the leaves. The bright red, pencil-shaped flowers are sparsely arranged on simple or branched inflorescences.

Aloe succotrina

Plants grow as robust, branched shrubs. The dull green leaves are carried erectly and are beautifully margined with short, white teeth. Inflorescences are elongated and carry dull red, pencil-shaped flowers along their lengths. When not in flower, *Aloe succotrina* can easily be confused with *A. arborescens*, but the dried leaves of *A. succotrina* are a distinctive purple colour.

THICKET (VALLEY BUSHVELD)

Thicket vegetation (sometimes broadly referred to as Valley Bushveld) is found in the valleys of most rivers that drain towards the coastline of the Eastern Cape and KwaZulu-Natal. Many of the woody plants that grow in these valleys and on the surrounding hills are spiny and, as they grow close together, they tend to form impenetrable thickets. Thicket-covered landscapes occur in parts of southern Africa that receive rainfall throughout the year, or mostly in summer. However, even though the rainfall can amount to several hundred millimetres, the valleys and surrounding hillocks are often in local rain shadows. They are therefore generally hot and dry (notably north-facing aspects), and succulents abound, particularly the spiny species. Thickets are generally not fire prone because of the abundance of fat-leaved species.

Parts of the distribution range of the genus *Aloe* are covered with sparse to impenetrable thickets. This applies especially to river valleys on the eastern and southern seaboard. These thickets are not as tall as forests, but they are very diverse in terms of species composition. On the road between Uitenhage and Jansenville *Aloe speciosa*, a typical thicket species, grows among dense bushes including succulent-stemmed cabbage trees, *Cussonia spicata*.

Aloe africana

Horticulturally, this single-stemmed species is the Eastern Cape equivalent of KwaZulu-Natal's *Aloe thraskii*, an exclusively coastal dune species. *A. africana* grows on the beach as well as inland. It is predominantly single-trunked, but branched specimens are common. Stems can grow tall and are crowned by a small rosette with strongly down-curved leaves. Tall candles of yellow-orange flowers appear at any time of the year, peaking in winter. The open flowers are characteristically upturned at the mouth.

Aloe greenii

These KwaZulu-Natal plants grow as small stemless specimens that are exceedingly prolific from the base. The white-mottled leaves are bright green, especially if grown in the shade, turning a deep brown that beautifully offsets the white flecks, if grown in full sun. Inflorescences are fairly tall and carry the dull red to dusty pink flowers in lax racemes. Plants are very easy to cultivate.

Aloe ciliaris

Plants grow as scraggly climbers that will, in time, form massive ball-shaped clumps if left without support for the thin stems. The leaves wrap around the stem, and the section around the stem is fringed with short, white hairs. The inflorescences are fairly short, but densely flowered. Flowers are pencil-shaped, bright red with yellow mouths. This is one of the easiest of all aloe species to grow; just stick stem cuttings into the ground where the plant is needed!

Aloe lineata var. lineata

These are medium-sized shrubs with a number of erect stems that are covered by the remains of dry, dead leaves. The leaves are a delightful bluish green colour and are adorned with numerous red, longitudinal lines, the character for which the species was named. Inflorescences are thick and stout and usually only the bottommost open flowers are not protected by big, tightly packed bracts. The colour of the open flowers varies from light orange to deep red.

Aloe pluridens

This typical thicket species has a very tall, smooth trunk, covered with the papery remains of dry leaves only just below the rosette. The leaves are fairly soft and strongly curved inward, towards the stem. They carry numerous light green to white, rather harmless marginal teeth. Inflorescences are very neat and carry closely packed, pencil-shaped flowers in a tidy arrangement.

Aloe speciosa

Plants grow as robust, single or branched specimens. Leaves are virtually unarmed, light blue and variously twisted. The large, fat inflorescences consist of hundreds upon hundreds of densely packed flowers, red in the bud stage and pure white when they open. It favours the mild climate of thicket vegetation, as well as the even milder, adjacent subtropical areas.

Aloe striata

Plants are stemless or short-stemmed. Leaves are very broad, boat-shaped and a beautiful light blue to sea green colour, with distinctive pinkish margins, but lack teeth. The small, bright orange flowers are arranged into dense, head-shaped inflorescences. It is a strikingly beautiful species, especially if planted en masse in a large garden. These plants are very easy to cultivate and offer gardeners few challenges.

Aloe tenuior

One of only a handful of scramblers in *Aloe*, it has thin, climbing stems that cannot support the weight of the small rosettes carried at their tips. The leaves vary from bluish green to light green. It flowers profusely and, during the summer flowering season, each rosette bears a number of small inflorescences, making it ideal for gardens in a mild climate. It also flourishes in more severe climates, where it dies down to ground level in winter and resprouts new stems in early spring.

TROPICAL, SUBTROPICAL AND AFROMONTANE FORESTS

A number of *Aloe* species favour the mainly climatically mild subtropical to tropical eastern parts, a region with some of the most magnificent natural scenery in southern Africa. Rain falls mainly in the hot, humid summer months, while the winters are dry, mild and pleasant. Many parts of the tropical to subtropical east coast have a higher rainfall than is normally associated with concentrations of aloes and succulents, but a multitude of succulents do occur in this region, which boasts a total flora of well over 6 000 species.

Of the two principal types of succulents (leaf and stem), stem succulents predominate in these eastern parts. A significant number of aloes, ranging from trees to miniature grass aloes, also occur here, further evidence that aloes can thrive in areas of high humidity.

Moving inland, and with increasing altitude, the climate becomes more temperate, and the tropical/subtropical coastal forest is replaced by grassland and small pockets of temperate or mountain forest, which is confined mainly to fire-protected gorges and valleys, especially along the Great Escarpment. These temperate (or Afromontane) forests are, however, rather impoverished as far as aloes are concerned.

Aloe barberae (right) towers over most of the trees in this forest patch near Louw's Creek in Mpumalanga. Almost by definition, not many aloes, which are sun-lovers, grow in forests, so some species that occur in the forest fringes and coastal forests, including those in which *A. rupestris* (opposite) is found, are also discussed here.

ALOES BY HABITAT

Aloe alooides (Skirt aloe)

Plants grow as large, single-stemmed specimens. The deeply channelled leaves are very long and recurved, almost reaching the ground. Inflorescences are tall and densely covered with small, yellowish brown flowers. The species is one of the few aloes that favours temperate forest margins, but it also ventures into adjacent savanna vegetation. It has a restricted geographical distribution range in Mpumalanga.

Aloe barberae (Tree aloe)

The tallest aloe, with specimens growing to 20 m. Trunks are smooth, with small to medium-sized rosettes perched at the tips of branches. Leaves are deeply channelled and armed with short, white teeth. Many-branched inflorescences are quite short, hardly protruding above leafy rosettes. Flowers are thick and fat, ranging in colour from deep orange to light salmon-pink. Regardless of colour, the open flowers are consistently a lighter shade than the buds and closed flowers.

Aloe rupestris

This is a typical, usually single-stemmed, tree aloe. It has large rosettes consisting of numerous leaves that are borne more or less horizontally to somewhat down-curved. Its beauty lies in its densely contracted inflorescences; although the flowers are a dull yellowish colour, the stamens are bright crimson-red, lighting up the entire inflorescence.

Aloe thraskii

Plants grow as large, single-stemmed specimens that can reach a height of up to 2 m. The deeply channelled leaves are large, boat-shaped and gracefully recurved towards the ground. The stem is often entirely clothed in a skirt of the remains of old, dry leaves. Leaf margins are armed with short, stout teeth. Inflorescences are quite short for such a robust plant and densely packed with small yellow flowers. At maturity, the stamens are far exserted, giving the inflorescence a feathery appearance.

GRASSLANDS OF THE HIGHVELD AND THE CENTRAL INTERIOR

Large parts of the interior and eastern Great Escarpment were once covered by vast rolling grasslands. Although their original extent has been severely reduced and fragmented by human activity, these remain among the most productive landscapes in the country and are intensely farmed, particularly for grain crops, timber trees and domestic livestock. The Witwatersrand (and most of Gauteng Province) is situated on the grasslands of the climatically severe Highveld, where the rain falls in summer and the winters are cold and dry.

Grasslands are characterized by sparse to dense stands of a great diversity of different grass species, sometimes interspersed with low-growing shrubs and herbaceous plants. The expansive, high-rainfall grasslands of Lesotho are dotted with impressive sandstone outcrops. This is one of the coldest areas in southern Africa.

Aloe ecklonis

Plants are solitary or grow as small clumps in sparse to dense grassland. Although it is a grass aloe, the leaves of *Aloe ecklonis* are much wider than the blades of the grasses among which it grows. Flowers are usually dull pink and arranged into short, dense, head-shaped inflorescences. These plants are adapted to tolerate very low temperatures, as well as fires. Because of their small size, grass aloes tend to be neglected by gardeners. However, the flowers of these species are usually larger than would be expected of such small rosettes. In order to grow them successfully they require somewhat more water in summer, but always a friable, well-drained soil mixture. (See also pages 43 and 46.)

Aloe graciliflora

Aloe graciliflora is sometimes regarded as a variant of, and so not distinguishable from, *A. davyana*, which in turn is sometimes formally treated as a variety of *A. greatheadii*. The rosettes are invariably solitary and produce a solitary or, rarely, once-branched inflorescence. The flowering time is a bit later than most species that grow on the Highveld. The flowers are a more intense red than those of *A. davyana*, with very obscure longitudinal stripes. This species is exceedingly cold hardy.

Aloe peglerae

Plants grow as small, mostly single, ball-shaped rosettes that hardly protrude above the dense grasses with which it is often associated. Leaves are incurved, short, chubby and sharp-tipped, with numerous dark brown teeth along the margins. Inflorescences are short and densely flowered. Flowers are red when in bud, opening to a dirty white colour, which is obscured by the exserted stamens. A beautiful species, but not easy to cultivate, and plants removed from the veld invariably die.

Aloe polyphylla

Plants grow as single rosettes or, rarely, as very large two-headed clumps. The leaves are reminiscent of those of some century plant (*Agave*) species and are carried in distinctive spirals. In mature specimens, the leaf margins are adorned with near-translucent, bony edges. The pinkish red flowers are carried in single or branched inflorescences. *Aloe polyphylla* is restricted (endemic) to the mountain kingdom of Lesotho, which is an indication of its specialized growing conditions.

Aloe pratensis

Plants grow as small, solitary rosettes that consist of numerous leaves borne close to the ground. Leaves are sharp-tipped and their margins are armed with numerous sharp teeth. The inflorescences are inverted cone-shapes and very large for the comparatively small plants. The flower buds tend to be densely packed and covered by large floral bracts. The open flowers are a pleasant deep orange to reddish colour.

Aloe striatula

Plants grow as robust, multi-stemmed shrubs that, over time, make neat, rounded specimens. The leaves are generally strongly recurved and widely spaced to reveal the distinctly lined leaf bases on the stems. The flowers are closely packed and a bright yellow colour. When grown close together, *Aloe striatula* plants make an effective boundary for *kraals* (animal enclosures) and are often grown for this purpose, especially in Lesotho.

Aloe suprafoliata

Plants commonly grow as medium-sized, rosulate specimens that may slowly produce many offshoots from the base. In young plants, the leaves are arranged into a single row. Leaf colour varies from dull green to a dusky, purplish green. Leaf margins are armed with small, white teeth. Red pencil-shaped flowers are arranged into neat elongated candles.

Aloe verecunda

Plants grow as short, multi-stemmed tufts in protected pockets in suitable microhabitats (often among rocks) throughout the grasslands. The leaves are typically grass-like and not very succulent. The flowers are quite large and arranged into small head-shaped clusters. This is one of the grass aloes that are characteristic of South Africa's interior grasslands.

Savanna regions are characterized by distinct grass and tree layers. In the dry winter season, the African grasslands, such as here, near the border between South Africa and Zimbabwe, turn yellow as deciduous trees shed their leaves. In the last rays of the setting sun, a flowering *Aloe angelica* towers over the landscape where it grows naturally.

SAVANNA (BUSHVELD)

Savanna, locally called bushveld – a landscape of grassy veld interspersed with trees and bushes – forms the iconic landscape image of Africa. The trees are solitary or in dense or sparse clumps, and often have flat-topped canopies, as in some *Acacia* (thorn trees) and *Albizia* (false thorns) species. In bushveld, the ground layer consists of an exceedingly diverse range of grass species, nearly 800 in southern Africa alone, of all descriptions and sizes. The bushveld regions typically receive summer rain, and their climate can best be described as mild and subtropical, although the severity of rather frequent cold spells in winter should not be underestimated! The densest concentration of different species of *Aloe* in the world is found in savanna, particularly around the town of Burgersfort in Mpumalanga, a region known as Sekhukhuneland.

Aloe vanrooyenii (left) and *A. castanea* (right) are typical of the species found in the savanna region.

Aloe affinis

This robust spotted (maculate) aloe, one of the larger members of the group, is mostly solitary. The light to mid-green leaves are contracted into strong, short-stemmed or stemless rosettes that hardly protrude above the dense grasses among which it grows. Leaf margins are armed with sharp, brown teeth. Inflorescences are much branched, giving the plant an overgrown appearance. Flowers are mostly deep red, with the unmistakable basal swelling of spotted aloe flowers.

Aloe angelica

Plants grow as solitary unbranched specimens that can reach a height of several metres. Leaves tend to be channelled and strongly recurved. Flowers are yellow, tinged with orange, and arranged into short, densely flowered inflorescences. The species is restricted to a small area in the Soutpansberg mountain range in Limpopo Province.

Aloe branddraaiensis

Plants grow as medium-sized rosettes that hardly ever produce suckers. The leaves are brownish green and mottled throughout with white flecks. Flowers are fairly small, bright red and arranged into head-shaped clusters on a much-branched inflorescence. This typical maculate aloe can be easily distinguished from its relatives because of the white flecks that become confluent into white longitudinal stripes, and the small clusters of bright red flowers.

Aloe burgersfortensis

This dainty maculate aloe seems to be more or less restricted to the vicinity of Burgersfort in Mpumalanga. Plants are fairly small, and the white-mottled leaves tend to be a dirty khaki colour. The flowers are dull pink and hardly distinguishable among the surrounding vegetation during the drab, dry flowering season. The inflorescences are loosely flowered, but grow taller than most aloes with similar sized rosettes.

Aloe candelabrum

Plants usually grow as single-trunked, or very rarely, branched, specimens that carry large, strongly recurved leaves in robust rosettes. The trunks are usually covered by a skirt of dry leaves. The leaves have large, smooth surfaces and are deeply channelled. The species, often regarded as synonymous with *Aloe ferox*, has dull red flowers arranged into thick, erect candles. Mature specimens of *A. candelabrum* tend to produce more candles per inflorescence than *A. ferox*, and its inflorescences often appear to be 'sunken' into a rosette. Some of the *A. candelabrum* plants that have been established alongside the N3 highway between Durban and Pietermaritzburg have developed into strikingly beautiful specimens.

Aloe castanea

These take the form of stout, branched trees with strong rosettes of long, dull green leaves perched at the tips of the branches. The inflorescences are variously curved to resemble cat's tails. The yellowish brown flowers are fairly small and tightly arranged on the inflorescences. The species was named for the chestnut brown nectar that is produced in copious amounts in the small, cup-shaped flowers. Within South Africa, it occurs in typical northern savannas.

Aloe chabaudii var. chabaudii

Plants grow as medium-sized specimens that slowly sucker from the base. The leaves are borne near-vertically, upcurved and, in the dry season, incurved. Leaf colour varies from typically bluish sea-green to light green. The flowers are small, ranging in colour from red to orange, and arranged into head-shaped to elongated, much branched inflorescences. This widespread species is easy to cultivate and is an excellent choice for mass planting.

Aloe cryptopoda

These stemless plants have robust rosettes of light to dark green, sharply tapering leaves. The leaf margins are armed with short, sharp-tipped, brown teeth. The inflorescences are shaped like inverted cones, while the individual flowers are fairly large and loosely packed on the inflorescence branches. Although it is a summer-rainfall species, it grows remarkably well in the Western Cape.

Aloe excelsa

Plants grow as tall, usually unbranched, specimens that can reach several metres. The boat-shaped leaves closely resemble those of *Aloe marlothii*, but tend to be somewhat shorter and more gracefully recurved. Inflorescences are branched and consist of several short, vertical to outwardly bent branches. The flowers, which range in colour from red through orange to yellow, are tightly packed in the inflorescences.

Aloe globuligemma

Plants grow as solitary rosettes that consist of numerous erect to apically recurved leaves. Bluish green leaves carry numerous whitish teeth along their margins. The inflorescence is horizontally branched, and the reddish white, distinctly club-shaped flowers point upwards. The flower mouths are hardly open, to prevent nectar from being diluted by rain, or concentrated through exposure to high levels of irradiance. The horizontal inflorescence branches make this a very distinctive aloe.

Aloe immaculata

A typical maculate aloe with medium-sized rosettes that can develop short stems. As the specific epithet 'immaculata' indicates, the leaves can be devoid of white, H-shaped spots commonly found on leaves of maculate aloes, while some forms of the species can have heavily mottled leaves. Leaf margins are typically armed with sharp, brown teeth. Inflorescences are usually branched, and carry pink to blood-red flowers with characteristic basal swellings in dense or loosely packed clusters.

Aloe lutescens

Plants grow as solitary, stemless specimens. The leaves are erect and incurved during times of drought. The inflorescences are bi-coloured, with the open flowers yellow and the buds red. Plants tend to remain considerably smaller than those of *Aloe cryptopoda* (see page 88) and *A. wickensii* (see page 95), both of which have a similar growth form and general appearance.

Aloe marlothii

This very variable species usually grows as a majestic specimen that consists of a strong rosette of densely prickled, boat-shaped leaves that are perched at the tips of robust stems. The stems are almost invariably clothed in the remains of dry leaves. The flowers are borne vertically on inflorescence branches that have horizontal side-branches.

Aloe parvibracteata

Plants sucker from the base to give rise to small to large clumps. The leaves are either light green and spotted with white flecks, or uniformly purple. Flower colour varies from light, dusty pink to orange. The species is very easy in cultivation and will thrive in virtually any type of soil. In cultivation, leaves tend not to die back, as they do in their natural habitat.

Aloe pretoriensis

A medium-sized, compact plant with fairly large rosettes, dense leaves and short stems. This species has one of the most striking inflorescences of all aloes, growing to several times the height of the plant, especially when young. It flowers when the plants are quite small, producing bright pinkish, almost perfectly pencil-shaped flowers. *Aloe pretoriensis* dislikes captivity, and transplants from nature invariably die. Even from seed, plants require a great deal of energy from the grower to keep them flourishing.

Aloe pruinosa

A typical spotted aloe; its long, snake-like leaves are cross-banded with densely arranged and confluent white blotches. The leaves are arranged into an open rosette precariously perched on a short stem that often bends and twists on the ground for part of its length, before turning upright to support the rosette in a perfectly erect position. The inflorescence is tall, much-branched and carries fairly dense clusters of dull pink flowers with white stripes. One of the most distinctive features is the dense layer of white powder (a waxy bloom) that covers all its above-ground parts, except the leaves. This species favours the somewhat milder environs of the eastern parts of southern KwaZulu-Natal, where it is restricted to a small area around Pietermaritzburg. It grows very fast and will flower within two to three years from seed.

Aloe spectabilis

Plants grow as robust, erect specimens that almost invariably remain unbranched. The leaves are large and boat-shaped, usually with teeth spread over both their upper and lower surfaces. Inflorescences are much-branched and consist of several near-erect racemes. Flowers vary from yellow to bright orange and tend to be arranged on one side of the inflorescence branches only. The species is sometimes included in *Aloe marlothii*.

Aloe spicata

Plants grow as medium-sized to large shrubs that consist of strong rosettes of numerous slender leaves, normally dull green but turning bright red in exposed positions. Inflorescences are densely packed with small, yellow flowers. Plants are easy to cultivate and this is one of the most common aloes of the savannas of Limpopo, Mpumalanga and northern KwaZulu-Natal.

Aloe transvaalensis

A nondescript, but variable, maculate aloe that flowers in summer and usually carries fairly insignificant dusty-looking, reddish pink flowers. This species of spotted aloe is an enigma; in the southern parts of its distribution range it seems sufficiently different from other forms of *Aloe zebrina* (in which it is sometimes included) to warrant segregate status as a species, but to the north it grades into *A. zebrina* proper. There are other forms from remote, and sometimes not so remote, parts of its range that are also quite different. For most of these, species status has been proposed in the past, but today all of them are treated as part of a variable (polymorphic) *A. zebrina*.

Aloe vanrooyenii

This is a recently described species from the savannas of the KwaZulu-Natal midlands. Rosettes are usually solitary, medium-sized and quite robust. Flower colour varies from red to deep orange. Unlike all the other maculate aloes that occur in the province, this species flowers in summer. A distinguishing feature is its very large fruits, which can attain the size of golf balls. A raceme densely covered in fully developed fruit invariably bends towards the ground. (See page 83.)

Aloe wickensii var. *lutea*

Plants are trunkless or very short-stemmed and always remain solitary. The leaves are lance-shaped and armed with short, rather pungent teeth on their margins. Inflorescences are single or branched, and consist of uniformly bright yellow flowers. The main horticultural attractions of this species, which is quite easy to cultivate, are its neat rosettes and yellow flowers.

NON-DISCRIMINATING ALOE SPECIES

Some *Aloe* species transcend the boundaries of the broad subdivisions of southern African vegetation types used here and could comfortably be included in any number of these regions. These species can easily withstand, indeed flourish in, a wide variety of climate, soil and topographical types. Almost without exception, they are also the most popular species in amenity and domestic horticulture.

The inflorescences of the East African *Aloe camperi* are head-shaped and produced in early summer, when few southern African species are in flower.

The bright, inverted cone-shaped inflorescences of *Aloe arborescens* are a well-known feature of gardens in South Africa's winter months.

Aloe arborescens

Plants grow as robust shrubs that, over time, form massive, multirosette-bearing clumps. The leaves are soft and easily damaged, typically rather narrow, sickle-shaped and adorned with stiff, but harmless teeth. Leaf colour varies from deep blue-green to a light yellowish green. The inflorescences are made up of a number of candles, each a strikingly beautiful inverted cone shape. Flower colour varies from bright yellow through all shades of orange and red to bright crimson. The species occurs in fynbos in the Western and southern Cape, in eastern coastal thickets, on the subtropical east coast, and in inland grassland and savanna. In cultivation, it is a popular plant for creating a 'living' fence or hedge.

Aloe camperi

This East African species certainly warrants being included here, given how widely it is cultivated in South Africa and across the world. Plants are robust rosettes that sucker from the base. The leaves are shiny green and adorned with white flecks, especially towards their bases. The inflorescences are fairly short and much branched. Flowers are yellow and orange and distinctly club-shaped. Plants thrive under virtually any growing conditions, ranging from deserts to the near-tropics.

Aloe davyana

These small, stemless plants sometimes form clonal clumps consisting of several heads. The triangular leaves are mostly short, the tips dying back for about one-third of their length. Leaf colour ranges from deep green to dark brown. Leaves are covered in irregular to H-shaped white spots. Inflorescences are single or multibranched, the latter carrying numerous cylindrical flower clusters, varying from dusty pink to deep red. Each flower has a distinct bulbous swelling at the base. Plants occur in the subtropical east coast, grassland and savanna.

Aloe grandidentata
Plants typically grow as very large clumps that consist of numerous rosettes. The leaves are very short, almost triangular in outline and beautifully mottled with white flecks. The inflorescences are branched and carry numerous club-shaped flowers, a character not usually associated with maculate aloes. In a dry garden, plants will in time form large cushion-like clumps, flowering freely from late winter into spring. This species occurs in typical grassland, savanna and karroid vegetation.

Aloe maculata

Plants are mostly small to medium-sized and stemless, but some forms will eventually produce a short trunk covered by the remains of dry leaves. The leaves are usually short and stubby, with distinct dried tips. Leaf surfaces are covered with a profusion of white, H-shaped spots. The inflorescences are made up of short, distinctly flat-topped or, more rarely, rounded candles (racemes). The flowers are quite large, slightly curved and have the distinct basal swellings typical of most spotted aloes. Flower colour varies from golden yellow through all shades of orange and red to crimson. The flowers are often covered with a dense layer of powdery bloom. The species is widespread, occuring in the fynbos, thicket, subtropical east coast, grassland and savanna regions.

Aloe vera

Aloe vera is the well known source of leaf sap extracts used in the medicinal and cosmetic industries. It probably originated from the Arabian Peninsula, but is now widely cultivated worldwide. Plants grow as medium-sized to large, stemless rosettes. The light green leaves have scattered white spots on both surfaces and short, stubby teeth on their margins. Leaves are quite soft and can easily be broken off or filleted. Branched inflorescences carry yellow flowers, sometimes tinged with orange, along an elongated stalk. The species is globally known as 'medicinal aloe' as a result of its soothing leaf juices, and is often grown as a ready-to-hand first aid treatment for burns and scratches. (See also Uses of Aloes, page 126.)

PART THREE
GARDENING WITH ALOES

In this large garden, the unmistakable bold outlines of *Aloe ferox* and *A. arborescens* are complemented by the carefully placed rocks and the variety of ground cover and companion shrubs.

*'There is ample evidence that any information on this subject
[the cultivation of aloes] will be welcomed by the public, and it behooves
the botanist to encourage this healthful appetite in every possible manner.'*

Reynolds in *Succulents for the amateur* by Brown et al. (1939): 147

Aloes are classic plants, naturally crafted by and for arid landscapes. Together with a range of related succulents, aloes are becoming increasingly popular with horticulturists and gardeners. They address a trend towards classical garden design, while still fitting in with more contemporary ideas of creating spare, uncluttered landscapes.

By observing a few simple rules, you can successfully grow these fleshy-leaved plants in climates ranging from mild and subtropical to harsh and arid. As they are well adapted to storing and conserving water in order to survive periods of drought, aloes offer low maintenance and easy care, whether they are planted in open beds or in containers.

Once gardeners have discovered the benefits of using succulents in general, and aloes in particular, they often find them irresistible. Switching to succulents provides welcome relief from the seasonal chore of replacing thick drifts of exhausted annuals. The cheerful red, yellow and gold of aloe flowers offsets their grey-green leaves and provides a colourful contrast, particularly to evergreen plants. Aloe shrubs, as well as the more tree-like aloe forms, can readily be used to give structure and form to a mixed garden.

Aloes, in particular, are perfectly at home in the African landscape and its prevalent climates. In addition, they look good when surrounded by earthy accessories created from local materials, like pieces of natural wood and rocks.

In order to grow aloes successfully, it is important to know the conditions under which they flourish in their natural habitats. This will give you an idea of how much exposure to the sun they like, the preferred soil type and conditions, the associated indicator species and vegetation, and a host of other useful biological and environmental information.

Aloe marlothii has pointed leaves covered with short, sharp teeth, which seem to echo the erect palisade-style metal spikes of the wall behind it.

GARDENING WITH **ALOES**

The yellow-flowered form of the multi-headed *Aloe arborescens* provides the visual background for a variety of densely cultivated indigenous plants.

GENERAL PRINCIPLES FOR GROWING ALOES

Although it is hard to generalize about the 'ideal' conditions and habitats in which to grow aloes, there are some guidelines that can be widely applied.

Rainfall and irrigation

Factors associated with rainfall, such as seasonality, amount and type, play an important role in, and affect the distribution ranges of, species of *Aloe*. For example, species that occur in the winter-rainfall region of southern Africa generally do not grow well in open beds in summer-rainfall regions, unless the beds are protected from unseasonal rainfall. The opposite is true of summer-rainfall species. Bear in mind that *Aloe* species from summer-dry areas need a very well-drained soil mixture, even if they are grown in other winter-rainfall areas, especially if the annual precipitation of their new home considerably exceeds the little received in their natural habitat.

Contrary to popular belief, aloes appreciate generous watering during the growing season, particularly during warm periods. Even during the resting phase, plants should be given a light watering to prevent unsightly die-back of the leaf tips. However, resist the temptation to drench them regularly at this time, as over-watering will almost certainly lead to root and/or stem rot. Watering is directly related to the soil mixture in which aloes are grown; plants grown in a friable, well-drained medium will demand more regular watering than those kept in a more clayey one.

As a rule of thumb in the southern hemisphere, aloes from summer-rainfall regions should be watered during months that contain an 'r' in their name. Watering should therefore be curtailed in May, June, July and August. In contrast, winter starts much earlier for winter-rainfall species and they can be given increased levels of irrigation from as early as March through to September.

Do not leave potted aloes standing in water for extended periods of time, as this can cause their roots to rot. For established aloes, regardless of whether the plants are grown in the open or in containers, overhead watering is best, as it flushes accumulated dust and soil particles from their leaves. However, in areas where the salt content of irrigation water is high, overhead watering will leave unsightly white crusts on the leaves and drip irrigation may be the best alternative.

Irrigating smaller aloes

Remember that large-growing plants are better able to store water than smaller ones. It is therefore preferable to continue to water small plants occasionally, even during the resting phase. All aloes will also benefit from an occasional misting, if for no other reason than to keep their leaves dust free. Misting also tends to raise the air humidity, which is something that aloes favour.

The robust leaves and dense, orange inflorescences of *Aloe thraskii* make for a striking feature in a coastal garden.

Aloe lineata var. *lineata*, a shrubby species originating from the Eastern Cape, works well in coastal gardens. The plants can grow about as tall as a grown man.

Air humidity

Generally, aloes grow better in atmospheres that have reasonable levels of humidity (the amount of moisture in the air). This applies particularly to thin- or flat-leaved species such as the grass aloes. During the non-rainy winter season, the air over much of the southern African interior is extremely dry. Coastal species that are transplanted to the interior tend to suffer, unless some additional overhead irrigation is supplied. In contrast, species from the interior transplanted to the coast generally do better in cultivation, for the coastal air is naturally much more humid.

Temperature

Temperature plays an important role in determining the geographical distribution ranges of plants, as well as determining which aloes can be used in gardens, and where. Most aloes flourish at high temperatures. However, in much of the inland distribution range of aloes, frost is common and this impacts adversely on coastal species transplanted to the interior.

Most aloes enter a vegetative resting, or dormant, phase during the cold winter months. (This applies much less to the few aloes indigenous to the Mediterranean region of the southwestern Cape where, among others,

Aloe plicatilis, *A. haemanthifolia*, *A. brevifolia* and *A. commixta* occur naturally.) However, regardless of the area from which the plants originate, almost all aloes can tolerate some light frost, especially if the soil in which they grow is relatively dry.

The majority of aloes flower in winter, making the flowers and inflorescences particularly vulnerable in areas that are subjected to very cold snaps, when they can suffer severe frost damage. Although aloe leaves tend to die back from the tips under these conditions, they usually recover very quickly when temperatures start to increase in spring. A general gardening rule is that roots are typically less cold-hardy than stems and leaves, and that, in frost-prone areas, aloes tend to be more tolerant of low temperatures if the soil is kept dry – they definitely don't like very cold, wet soil.

Generally speaking, temperatures below about -5 to -7°C will severely damage or kill most aloes. The few species that will survive beyond these temperatures include *Aloe davyana*, *A. striatula*, *A. polyphylla* and *A. transvaalensis*, whose natural habitats straddle the climatically severe Highveld or adjacent high-altitude areas. These species can survive low night temperatures for extended periods, while often the temperature the following day will be up to 20°C higher, with no apparent ill effect to the plants. Significantly, the majority of these species flower in summer (or at least not in mid-winter).

The medicinally important *Aloe vera* is widely grown all over the world. Here, it looks very much at home in a hotel garden near the town of Tavira, on Portugal's southerly Algarve coast.

Selecting cold-hardy forms

If you want to garden with aloes in very cold areas, make sure you either cultivate species naturally confined to such areas, or select forms of more widespread species that originate from the coldest parts of their range.

In species whose natural distribution covers a wide geographic range, several genetic forms best adapted to local environmental conditions may develop through natural selection over long periods of time. If some parts of a particular species' range are colder in winter than elsewhere in its range, plants originating from the colder parts are likely to be more cold-tolerant.

Known as ecotypes, these local genetic forms are often not morphologically distinct from other members of a species, hence they are usually not formally recognized by scientific names.

> ### Frost tolerance of selected aloes
> The list below is simply a guideline as to how well individual species can tolerate varying degrees of frost. The plants were selected based on cultivation on the Highveld.

Frost tender (to 0°C)
Aloe africana
Aloe brevifolia
Aloe plicatilis
Aloe thraskii

Light frost (to -4°C)
Aloe aculeata
Aloe barberae
Aloe ciliaris
Aloe comptonii
Aloe cryptopoda
Aloe excelsa
Aloe dichotoma
Aloe haemanthifolia
Aloe pillansii
Aloe ramosissima
Aloe speciosa

Mild frost (to -7°C)
Aloe chabaudii
Aloe ciliaris
Aloe cooperi
Aloe ferox
Aloe fosteri
Aloe globuligemma
Aloe maculata
Aloe marlothii
Aloe microstigma
Aloe petricola
Aloe pretoriensis
Aloe rupestris
Aloe spicata
Aloe striata
Aloe thompsoniae
Aloe vanbalenii

Severe frost (lower than -7°C)
Aloe arborescens
Aloe aristata
Aloe boylei
Aloe davyana
Aloe ecklonis
Aloe grandidentata
Aloe kraussii
Aloe lutescens (some forms)
Aloe mutabilis (some forms)
Aloe parvibracteata
Aloe polyphylla
Aloe reitzii
Aloe striatula
Aloe tenuior
Aloe transvaalensis
Aloe verecunda

Light intensity

Almost without exception all species of *Aloe* need ample, bright light, especially as mature plants. However, this does not necessarily mean exposure to the often harsh rays of *direct* sunlight. Many species, certainly as juvenile plants when their leaves are soft and their roots poorly developed, prefer shady conditions, with filtered – but still bright – sunlight. In their natural habitats this type of shading is provided by nurse plants, such as grasses, small karoo bushes or open-canopy trees. Even leaf litter can fulfil this role, particularly for seedlings.

The leaf tips of some species will die back regardless of the intensity of light to which they are subjected. The summer-rainfall maculate, or spotted, aloes are a good example of this phenomenon. However, their leaf dieback usually results from a combination of factors: low water resources during the resting phase; exposure to extreme temperatures; and high light intensity, which can occur when the loss of shading canopies in winter-deciduous trees suddenly allows considerably more light to reach previously shaded aloes.

If you keep your plants under the protection of glass in the resting phase, or during protracted cold, wet periods, you should continue to protect them from direct sunlight (and even rain) for the first few weeks once they are ready to be re-introduced to sunny positions – otherwise they are sure to get scorched.

Plants usually indicate clearly when they receive too much insolation (exposure to the sun), as their leaves turn a darker green, and ultimately a reddish

Aloe camperi, from East Africa, does well in full sun and partial shade, as here in the Desert Garden, Arizona; it flowers from spring to early summer.

Aloe pictifolia, a small cliff-dweller from the Eastern Cape, thrives in a dry wall rockery. As a rule of thumb, aloes love sun.

Aloe arborescens, a stalwart of aloe and succulent gardens, has a variable flowering season. This cultivar flowers in mid-summer.

brown. Although this may look attractive for a short time, it is a sign that the leaves are under stress. If this happens to plants in containers, they should be moved under cover, or to where they will receive filtered sunlight only.

A few species actually thrive in deep shade as adults, notably *Aloe suffulta*. Others, such as *A. davyana*, are able to cope with either shade or open, sunny positions.

Soil and drainage

Aloes are generally not fussy about the soil in which they are grown. The grass aloes are the exceptions: they tend to prefer an airy, very friable mixture with reasonable amounts of decomposed organic material (coarse compost) added to the soil. A very important aspect of soil type is its ability to drain well, while still retaining some water. Aloes cannot tolerate wet feet (i.e. soil with poor drainage). There are exceptions, however: in the Drakensberg mountains of the eastern Free State, *Aloe maculata* has been observed growing in very muddy soil that is virtually water-logged for the entire summer rainy season, with no apparent ill effects to the plants.

In open garden beds it can be difficult to ensure adequate drainage at all times, especially if you are gardening in clayey soils. Under such conditions, a reasonable quantity of sharp sand should be added to the hole prepared for the plants. Before planting a new aloe, the soil should be dug out of the hole and then mixed with copious amounts of sand and, ideally, some compost.

In this garden the yellow-flowered *Aloe marlothii* and the red-flowered *A. arborescens* are combined to great visual effect in an informal bed that glows with colour and life.

ALOES AND GARDENING STYLES

A certain playful attitude to our gardens and a sense of eclecticism will often free up our creativity and allow us to explore the use of plants we may not previously have considered using. However, regardless of where we live and the style of garden to which we aspire, the best possible gardening practice is to endeavour to work in harmony with nature.

When establishing a new garden, whether large or small, private or public, or implementing a garden 'makeover', it is possible to draw on a range of different elements and styles. These days, the single most important guiding principle is to aim for waterwise gardening; this is where aloes come into their own. The low-maintenance aspects of gardening with aloes are obvious: their strategic advantage is the nearly improper levels of neglect they will tolerate while still looking their best. They are durable, virtually indestructible, low maintenance plants – an important factor in today's economic climate.

If aloes and rocks are wisely combined, regular gardening chores can generally be kept to a minimum. Remember that for a garden to be a delight, it must also be practical. This means the gardener must have time to enjoy his or her handiwork. Otherwise, what is the point of gardening?

To maximize their natural visual impact, aloes can be planted in strategic places in the garden where they can be easily seen and enjoyed. Alternatively, they can be tucked away in unexpected corners where they become exciting finds. Aloes can be put to good effect in most garden styles. They show up to best advantage when they are given the freedom to develop into large, robust specimen plants; these often provide the most effective, eye-catching, focal points of a garden – especially when they flower prolifically, typically in winter.

For example, try planting distinctively bold aloes en masse to achieve a spectacular 'wall' of red at a time when most other plants are not flowering. These bursts of vibrant colour in winter express the independent spirit of aloe gardeners, and provide sumptuous viewing for onlookers. Aloes are equally striking in restrained and formal settings, such as in urn-shaped pots on Italianate verandas, or in informal gardens that have soft, flowing lines around winding, paved paths.

A particular advantage that aloes have over less robust herbaceous plants, which cry out for shade and irrigation in the midday heat, is that, as succulents, aloes are mostly unaffected by hot conditions. You can combat the rather tired look that many non-succulents take on during long, hot summer days, by replacing them with sun-loving aloes and related plants, creating an illusion of cool tranquillity.

Landscape developers, horticulturalists and home gardeners all have the advantage of being able to select from a growing range of aloes to suit diverse garden settings, whatever the style or climate. Gardeners need to experiment with different gardening styles and plants to see what works for them. Rather than gardening exclusively with aloes and succulents, a harmonious blend can be obtained by combining succulents with water-wise, non-succulent species. Thanks to the wide range of aloes and their sizes and shapes, they can be used in a multitude of landscape settings.

> ## Why aloes make good garden plants:
> - They enable the gardener to choose from a rich variety of growth forms.
> - They make excellent accent plants in most settings.
> - Aloes come in variants that are suited to almost any climate, including a surprising number that will grow under subtropical conditions.
> - Their leaves can be highly ornamental, offering a range of different textures and patterns – from spotted leaves to those with multiple teeth.
> - Aloes produce a seasonal flowering display of vibrant colour that outlasts that of many other plants.
> - Most *Aloe* species flower profusely.
> - The winter months, when many other plants are dormant, are the prime flowering season for most aloes.
> - Large, mature plants can be established with ease.
> - Well-grown, healthy plants are relatively pest-resistant.
> - Once established, aloe plants require little maintenance.
> - Aloes generally produce negligible leaf litter.

You can add drama to a garden by interplanting aloes with tough, but exotic, species like cordylines, yuccas and flaxes, which have smooth-edged, flexible leaves. Unlike aloes, the dead leaves do not have to remain on the plants to make them look natural.

Aloes are ideal for planting in well-drained sites where other plants might struggle, such as in a rock garden. The plants can be anchored with rocks and the spaces around the stem filled in with gravel.

The strong lines of aloes and succulents make them perfect for use in combination with stark features, such as rocks or boulders, walls and periphery fencing. Despite the plants themselves having harsh lines, they are excellent for softening natural or man-made structures that might otherwise be dominating. They also work well in garden projects that require layering. For example, if parallel, horizontal lines are required, an avenue of ball-shaped *Aloe arborescens* shrubs of near-uniform height, interspersed with *Aloe barberae*, will offer glimpses into the landscape beyond the hedge.

Aloe comptonii, a cliff-dweller in its natural habitat, is perfect for planting in an elevated garden rockery.

Gardening in a way that enhances, rather than damages, the environment includes conserving water and ensuring effective nutrient recycling through composting. Gardening with aloes and succulents also helps to attract nature back to our gardens, as aloes offer copious amounts of nectar and pollen for visiting birds and insects. With this in mind, it makes sense to gradually remove plants that impact on natural resources and replace them with plants that are environmentally friendly and easier to maintain.

When selecting aloes for your garden, it is important to choose species that are known to flourish in your particular area and local climatic conditions. *Aloe chabaudii* is an example of a species that can flourish in subtropical and even tropical areas if it is planted in soil with reasonable drainage. En masse, the blue-green colour of its leaves offers an austere, other-worldly charm and, when in flower, the dainty heads of the slightly curved, pencil-shaped flowers drift like a soft mist above the rosettes.

Above: Although some forms of *Aloe striatula*, a southern African grassland specialist, are shy to flower in cultivation, this clone is very much at home in the mild Mediterranean climate of Monaco's Jardin Exotique. *Agave attenuata*, from Mexico, grows in the foreground (on the right).
Inset: *Aloe striatula's* inwardly curved yellow flowers are densely clustered into short inflorescences.
Left: Aloes en masse make for a colourful display when they flower; these *Aloe striata* grow alongside the old national road between Uitenhage and Port Elizabeth.

GARDENING WITH **ALOES** 113

A hybrid between *Aloe humilis* and *A. arborescens*, widely referred to as *Aloe* x *spinosissima*, grows happily in an ornate container in a large garden in Tuscany, Italy.

GROWING ALOES IN CONTAINERS

Pot plants often substitute for cut flowers, bringing instant cheer to a house, whether for interior or exterior decoration. But few survive beyond their first flowering season, after which they are consigned to the compost heap. With aloes, this need not be the case. In nature, aloes often grow in tight rock crevices, making them perfectly adapted for container growth. An aloe in full bloom makes a particularly decorative statement in the dreary winter months, whether indoors or outdoors. Even after they have finished flowering, their foliage can brighten an otherwise dull corner. If you grow your aloes in containers, you can spice up your surroundings by simply rearranging the pots.

Although container plants that are kept indoors are protected from harsh conditions, this also means that they are generally kept away from life-giving natural light, nourishing rainwater and organically recycled plant food. To ensure that pot plants remain healthy, certain routines are necessary. Make sure the plants are kept clean: this can be done using a fine spray of water, or by carefully wiping the leaves of large-growing species with a soft cloth and tepid water to remove any accumulated dust. Regular feeding will compensate for the lack of organic food. If the containers are small enough to move outside, allow your potted plants to enjoy a shower of rain from time to time.

In time, even medium-sized plants will form short trunks clothed in the remains of dried leaves. It is not necessary to remove these leaves, as they tend to give the plant a more natural look, but they can harbour parasitic insects which can be difficult to control indoors without using chemicals. One response is to introduce predatory insects to feast on the insect infestations – although regularly removing the spent leaves is perhaps a more obvious option.

When to repot?

When the roots of plants start to appear from the drainage holes, it is a sign that they are ready for repotting. Aloes grown in containers can be repotted at almost any time of year, but preferably not when the plants are in flower or have started to produce inflorescences. The shock of a severe disturbance of the roots of a flowering aloe will almost certainly result in the loss of already open flowers or a complete cessation of flowering for the season.

Repotting aloes

Aloes are perennials. This means that they can be kept in containers virtually indefinitely. However, as with all perennials, it is necessary to repot them from time to time to give them room to grow.

Carefully tap the overgrown aloe out of its pot, then carefully tease apart the roots and remove as much 'dead' soil as possible. Inspect the root ball to ensure that the roots are healthy and free of root parasites. Using a pair of sharp secateurs or garden scissors, cleanly sever any damaged or diseased roots.

Make sure the new container has sufficient drainage holes and that these are big enough and unblocked. You can prevent drainage holes from blocking up as a result of soil compaction by covering them with pieces of broken crockery, or material such as gravel, gauze, even undecomposed tree leaves. (It is not necessary to place a layer of gravel at the bottom of the pot: this just takes up space that could be used for additional soil.) Cover the bottom of the pot with an appropriate soil mixture (see below), to a depth that will ensure the plant will be buried to the same level as it was in the old pot. Suspend the plant in its new pot, with its roots dangling naturally, and fill the pot to within about one centimetre of the rim. Tap the sides of the pot to allow soil to settle around the exposed root ball. Press down the soil around the stem or rosette of the plant to prevent it from simply washing down into the pot with the first watering. Use your hands to do this so as not to compact the soil too much. The freshly added soil should be just moist, not too dry and certainly not too wet. Moist soil will stimulate the roots to grow into the fresh soil. Wait about one week before watering the newly planted aloe.

To avoid slow-down in the growth of a potted aloe, you can transplant it by lifting the entire plant from the old pot, retaining as much soil as possible around the root ball, and carefully placing it in a larger container before packing it in with fresh soil. Although not ideal, it is not particularly detrimental to retain the old, 'dead' soil, as the aloes's roots will soon find their way into the fresh soil. This process reduces the shock of repotting, and plants will quickly recover and produce new, vigorous growth.

Soil mixtures

Very few garden soils should be used directly and without additives, such as grit and compost, in a container. Garden soils simply do not have the physical properties and ability to provide the roots with sufficient air, water and plant food to ensure healthy plant growth in a restricted space.

Potted plants are entirely dependant on the soil mixture you prepare for their nutrition. For container cultivation, a soil mixture consisting of equal parts of well-rotted compost (purchased or prepared from garden clippings), sharp river sand and good garden soil will yield excellent results. Mix these components thoroughly and sift them to remove large soil clods and create a mixture of equal consistency. Ideally, soil should be devoid of weed seed, excessive amounts of salts and pathogenic fungus spores. It is often preferable to purchase balanced pot-plant soil that has been prepared with pot culture in mind. It is important to supplement the nutrients in the soil by regularly feeding your plants. Aloes respond very favourably to seaweed extract, which should be applied as an overhead drenching about once a month in the growing season.

In general, succulents should not be excessively pampered – they should be encouraged to look natural. Applying fertilizer too often or too generously will make them appear pumped up and decidedly unnatural. The proportions of nitrogen, phosphorous and potassium in a fertilizer are shown on the container as N:P:K values. For aloes, the first figure (indicating the nitrogen level), should be low in comparison to the other two, as too much nitrogen can lead to soft, undesirable growth.

Choosing a container

Always choose a pot that will accommodate the eventual size of the plant, specially if you are planting a near-mature specimen. With the exception of some small and miniature species, aloes mostly require a large pot to attain full blooming maturity as speedily as possible.

To achieve the best appearance, the pot or container should be only slightly larger than the final flowering size of the plant and its root system. Since aloes do not have long tap roots (their roots are soft, succulent and fibrous), if there is any doubt about the eventual size of the mature plant, it is probably better to use a pot that will be slightly too small.

Having provided adequate space for root and rosette development, you can provide the ideal conditions for potted aloes to flower as regularly as possible. For most aloes, this is once a year, but there are exceptions.

Although aloes are hardy and will grow and flourish in almost any type of container, your choice of pot will have some bearing on the appearance of the plants. Apart from aesthetic considerations, you may want to think about mobility (how easy it is to move the pot); environmental conditions (how quickly the soil will dry out between irrigations); and the material from which the pot is made (clay, ceramic, plastic, etc). Pots can be chosen to convey different themes: for example, an African savanna feel can be achieved with a series of different-sized aloes in calabash-style clay pots.

However, while it is undeniable that aloes look remarkably good in natural clay pots, bear in mind that, when filled with soil, clay pots can be exceedingly heavy and difficult to move around. Also, since clay is porous, the soil will dry out rapidly. As a rule of thumb, unglazed clay pots dry out twice as fast as plastic pots.

Plastic pots are widely available in a range of attractive finishes, and at fairly reasonable prices. They are non-porous and water-tight, so it is important to ensure that the drainage holes do not get clogged up.

Although containers without drainage holes do not require drip trays, it is best not to use them for aloes as it is easy to over-water plants grown in them, which invariably leads to the demise of the plants. If you particularly like a container that does not have drainage holes, drill some holes in the bottom of the pot and place it on a drip tray made of similar material if the plant is to be kept indoors.

Where to place plant containers

The proximity of containers to heaters, lamps and air-conditioners will have an effect on plant growth. Ensure your pot plants receive adequate air movement (ventilation), sufficient light and reasonable levels of air moisture (humidity). It is a myth that all aloes prefer dry heat. In fact, the opposite is true of many aloes, except perhaps those that come from very arid deserts. Aloes grown indoors will benefit from having their leaves misted occasionally.

The leaves of all indoor plants, including aloes, become accustomed to slightly lower light intensities. When they are briefly placed outdoors, to benefit from a downpour, for example, it is important not to expose them to full and direct sunlight for prolonged periods, as this can scorch the leaves, resulting in unsightly spots. In extreme cases, the plants may actually shrivel and die. An outdoor recovery period should be introduced slowly, by periodically exposing plants to increased levels of light intensity.

Aloes are offered for sale in many parts of the world. This potted hybrid of *Aloe perfoliata* was on offer in a department store in downtown Yokohama, Japan.

The dainty creeper, *Aloe tenuior*, which can flower at any time of the year in a mild climate, contributes colour to a mixed bed in a suburban garden.

GROWING AND PROPAGATING ALOES

GROWING ALOES FROM SEED

Aloe seed should be harvested as soon as the large, fat capsules become dry and split open to expose the usually small, black seeds. If you require aloe seeds in large quantities, you can remove an entire inflorescence as soon as the lowest capsules start to split open. Stand the inflorescence upside down in the shade on a large sheet of paper onto which the seeds will be released. If insect infestations of the capsules are likely, cut the inflorescence into manageable lengths and enclose them in a paper bag to which an insecticide or fungicide has been added. Never use a plastic bag as any moisture remaining in the inflorescence will cause condensation, which will inevitably lead to fungal growth on the seeds. Ripening capsules are sometimes attacked by insects that lay their eggs in them, leading to the destruction of large numbers of seeds. It helps to spray the developing capsules with a systemic insecticide to prevent large-scale damage.

Aloe seeds are predominantly adapted to wind dispersal. This is obvious, given the presence of small wings on the seeds, and the fact that the flowers (and therefore, ultimately, the capsules and seeds) are usually borne on tall inflorescences that are well exposed to air movement. Furthermore, ripe capsules tend to turn upwards, making it easy for seeds to be released from the cup-shaped capsules. If the ripe capsules did not turn upwards, the dispersed seeds would either end up in the crown of the mother plant or very close to it, making for fierce competition among the emerging seedlings.

It is not necessary to remove the wings before sowing the seed. Even when the wings are prominent, as in *Aloe variegata*, they do not stimulate the development of fungal growth as may be the case in other, unrelated species.

The leaves of some Aloe seedlings, such as *Aloe striata* (left) and *A. petricola* (right), are distinctly two-ranked, but this will change into a rosulate arrangement as the plants mature.

Germinating aloe seeds

For germinating aloe seed, a mixture containing equal parts of coarse river sand and sifted compost is suitable. The optional addition of some wood ash (sifted from the remains of the fire from your last braai) provides a useful source of essential trace elements. The size of the seedling tray is not critical, but if it is unlikely that the seedlings will be planted into individual pots as soon as they are ready for repotting, it is better to sow the seeds in a tray that will allow for some root-run.

Aloe seeds are generally large enough to allow them to be placed individually on the soil surface in a seedling tray, but this is a rather tedious way of sowing them. It is much easier to 'broadcast' them onto the soil surface, covering them with a very thin layer (1–2 mm, i.e. more or less their own depth) of soil or pure sand after sowing, followed by a final layer of pea-sized gravel spread over the soil surface. (It is not necessary to make holes in the soil when sowing seeds individually.)

Suitable microhabitats, such as those created by the shading effects of soil clods or gravel, are beneficial, indeed essential, for the germination of aloe seed. Since the roots of aloe seedlings are extremely flimsy and not at all comparable to the size of the first above-ground leaves, the layer of gravel provides support for the emerging seedlings. It also prevents soil from splashing onto the plants once watering from above is started.

In their natural habitats, aloe seeds tend to germinate under the protection of surrounding plants, also referred to as companion or 'nurse' plants. The microclimate beneath these plants is often more amenable to germination than out in the open, as it offers moisture and shade, and the surrounding plants help to protect the tender young leaves from becoming sunburnt. To achieve the same effect, keep seedling trays in the shade, away from the intense rays of the sun. The seeds of most species should start germinating within one week of sowing.

Senecio rowleyanus, commonly known as string-of-pearls, spends its entire life cycle in the dappled shade of companion, or nurse, plants. Here the wiry stems of the species sprawl over the branches of a small karroid shrublet. However, *Aloe* seedlings will eventually outgrow the nurse plants under which they develop.

Aloe seeds do not require a resting period and should be sown fresh. However, this does not mean that seeds that are a few seasons old will not germinate. A certain degree of staggered viability among seeds in a single batch has been observed and, often, at least some seed from an old batch will still germinate, albeit inconsistently and in much reduced numbers.

In the southern hemisphere, seed should be sown any time from August to February, when the night temperatures tend to remain above 10°C.

Seedlings benefit from the addition of a weak solution of liquid fertilizer (particularly one derived from a seaweed extract) to the irrigation water. This helps them grow stronger and faster, and they tend to remain healthy. Do not be tempted to boost seedling growth with full-strength fertilizer as this can burn their young, soft leaves and roots.

To avoid disturbing the developing roots, the seedling tray should be watered from underneath until the seedlings are firmly established. You can do this by standing the seedling tray in a shallow drip tray to which water has been added. After a few minutes, the soil in the seedling tray will be thoroughly soaked. Adding a water-soluble fungicide to the water will prevent the damping-off of the seedlings. The soil in the seedling tray should not be allowed to dry out completely: aloe seedlings thrive in moist, but not wet, conditions.

Once all, or most, of the seedlings have appeared, watering can be scaled down somewhat. But bear in mind that, although most species of aloe come from mild to warm summer-rainfall areas, it is a fallacy that aloes enjoy extreme, desert-like conditions, so don't under-water your seedlings.

Aloe seedlings should be ready to be transplanted when three true leaves have formed. This can normally be done after a year or less, depending on the species in question. Seedlings can be planted into individual pots or, if they have grown really fast, directly into open beds. However, they can also stay in the seedling trays for quite some time. It is not necessary to transplant them as soon as they appear to be crowded. The apparent root competition induced among seedlings growing densely together seems to encourage faster rosette development.

Cross-pollinating aloes

With few exceptions, aloes are 'obligate outbreeders'. This means that at least two plants, ideally from different parentage, are required to produce viable seed. To effect cross-pollination between two plants is not difficult: essentially, fresh pollen has to be transferred from the anthers of one plant to receptive stigmas on the flowers of another plant.

Aloes are protandric, which means that their pollen ripens before the stigmas become receptive. Therefore, pollen has to be transferred from the youngest open flowers on one plant to the stigmas at the tips of the styles of older flowers on another plant. By now these styles and stigmas should be protruding beyond the mouths of the older flowers.

The jellybean-shaped flowers of *Aloe ortholopha* are borne vertically on horizontal inflorescences. This aloe is restricted to the Great Dyke in Zimbabwe and presents some challenges if it is to be grown successfully outside its natural distribution range.

TAKING CUTTINGS

Growing aloes from cuttings is perhaps the easiest method of establishing new plants. Hardwood cuttings of woody, arborescent shrubs can take up to nine or 12 months to strike root; in contrast, aloes take only a few weeks which, for the gardener, means instant foliage in an intended spot. One attraction of growing aloes from cuttings is that new plants established in this way can start off as quite large plants, and will quickly come into flower.

To take a cutting, use a sharp knife or pair of secateurs to remove a rosette-bearing stem from the parent plant. Take the cutting from low down on the parent plant so as not to leave an unsightly scar or a rosetteless stump. The cutting can be allowed to dry in the shade for a few days before it is planted, but in some species, for example, *Aloe arborescens*, this is not necessary. Dusting the cut edges with commercially available flower of sulphur will help to prevent rot, while an application of rooting hormone powder will stimulate the formation of roots, although most aloes will easily form roots without the external application of any stimulants.

Cuttings from a species with naturally flimsy leaves, such as some of the miniature grass aloes, don't necessarily need a drying-off period, but these slips should be planted immediately and only watered after a few days have elapsed, and then only lightly, until they are established and show signs of growth. They will benefit from initial protection from harsh sunlight, which will prevent the leaves from turning yellow or brown while the plants struggle for survival without roots.

Neither the rosettes nor the lower leaves of aloe cuttings should be buried when they are planted; the plants should appear to be 'sitting' on the ground.

Aloe plants can be established very quickly from cuttings. This light orange flower form of *Aloe arborescens* flowered within one season of being planted as a large stem cutting.

PROPAGATING ALOES THROUGH DIVISION

Clump-forming species, such as *Aloe aristata*, some of the maculate aloes that readily form offshoots, and the smaller Madagascan aloes, can all be multiplied by splitting a clump into several individual plants. Minor plants removed from the parent plant usually already have developed roots and can be planted immediately into their own pots or in the garden. Plants that already have roots become established quicker than those that don't.

If the offshoots don't have roots, the wounds should first be allowed to dry by leaving the slips in the shade for up to five days. They can then be planted by 'sitting' them on the ground. New roots will quickly grow into the soil mixture, resulting in a flourishing new specimen that will soon begin to form more offshoots. However, it is not uncommon for such an offshoot to lose its lower leaves, so do not be alarmed if this happens.

Where an *Aloe barberae* stem has been damaged, it will often produce offsets that can be removed and grown on.

If a growing tip is damaged, such as in this *Aloe peglerae*, the plant will often produce several heads from the apex.

PESTS AND DISEASES

The first rule for having a pest-free and disease-free collection of aloes is to grow them well, as a healthy plant is far more resistant to attacks from the natural enemies of these generally very hardy plants. It is wise to pay attention to where aloes are placed, regardless of whether they are grown in open beds or in containers. For example, plants that are not exposed to sufficient levels of irradiance will generally be grown too soft and will be particularly prone to attacks by insects and arachnids. Most pests and diseases that attack aloes can be easily controlled with the application of a systemic pesticide or fungicide.

However, regardless of how much attention you pay to their growing conditions, some aloe species simply attract more pests than others. *Aloe marlothii* is a case in point; in a garden, and even in its natural habitat, it seems to be prone to every conceivable malady. In such a case, it is wise to grow plants from seed which has been obtained from a population of healthy and apparently resistant plants.

Black rust is clearly visible on the leaves of this specimen of *Aloe marlothii*. If left unchecked, this disease can lead to the demise of a plant.

Inflorescences, such as this of *Aloe variegata*, are often attacked by aphids which, in turn, are controlled by predatory insects, such as ladybirds.

Above: This cancerous growth on the stem of a specimen of *Aloe arborescens* has been caused by a mite infestation. After successful treatment, the surviving plantlets can be grown on as healthy individuals.

Above and right: In their natural habitats, aloes are able to cope with, and protect themselves against, destructive insect infestations. Here, white scale insect has attacked, but not destroyed, some leaves of *Aloe littoralis* (above) and *A. cryptopoda* (right) plants, while most of the leaves have remained unaffected. Note how the white scale on *A. littoralis* resembles the white flecks on the lower parts of the leaves.

PART FOUR
USES OF ALOES

When cut, *Aloe marlothii* leaves produce bitter yellow sap that is processed for medicinal use.

The pulp of *Aloe ferox* leaves can be made into traditional *konfyt* (preserve).

Aloes have been used for medicinal, cosmetic and other cultural purposes since time immemorial. The healing properties of *Aloe vera*, in particular, which likely originates from the Arabian Peninsula, were known to many early civilizations throughout the Mediterranean basin and beyond. Today, the centre of *A. vera* cultivation is Central America and the southern USA, but South Africa's *A. ferox* is making inroads into the traditional pharmaceutical and cosmetic markets that until now have been dominated by *A. vera*.

Locally, the commercial extraction of components of the sap and leaf pulp of *Aloe ferox* for use in a variety of products is confined largely to the Albertinia-Mossel Bay area of the southern Cape. Harvesting mostly follows traditional methods, with only a few older leaves being taken from each plant at a time. The leaf of the aloe plant yields two main components: bitter, yellow sap, which bleeds from just below the skin; and the inner, fleshy, mucilaginous gel. The bitter sap is drained from the fresh leaves, then heated until it chemically reduces and becomes concentrated. The sap is then allowed to cool and solidify into a block from which crystals or powder are processed. Once the leaves have yielded their sap they are washed, sliced and minced to obtain the aloe gel, which is then further processed into various end products.

In order to gather the yellow to bright orange exudate that bleeds from *Aloe ferox* leaves, they are traditionally stacked in a circular pile over a tarpaulin placed in a hollow in the ground. The cut ends face inwards to allow the juices slowly to collect in the depression (left). The resulting sap is heated in drums over a fire and the concentrated dark brown, lumpy crystals derived from this process are sold as 'Cape aloes'. Once the sap has been removed, the leaves are often discarded and left to decompose (right).

Aloe vera and *A. ferox* are the two best-known aloe species with medicinal and cosmetic applications, but *A. davyana* and *A. arborescens* display similar properties and are often used by those seeking alternatives to conventional medicines. In some rural areas, the dried leaves of the mountain aloe or *bergaalwyn*, *A. marlothii*, are ground to yield a snuff.

Traditional medicinal uses

Traditional medicine is practised throughout southern Africa, particularly in rural communities where access to medical facilities is limited and financial constraints mean making the most of what nature has provided. *Aloe ferox* is most often used, but other members of the aloe family may be utilized in areas where *A. ferox* does not occur naturally.

Fresh leaves are boiled to make a decoction which is taken as a laxative. Dried crystals of leaf sap can be taken orally as a laxative and are also used to treat arthritis, while the fresh bitter sap is used to treat conjunctivitis and sinusitis. Leaf gel is applied externally as a remedy for skin irritations, cuts or bruises.

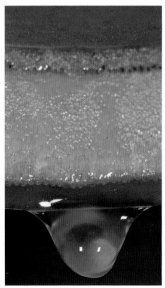 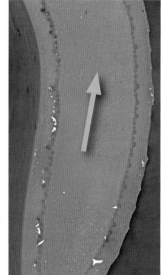

Aloe sap (left) and fleshy inner leaf gel (right).

Traditional medicines can have adverse effects if they are prepared or used without adequate supervision and knowledge. Due to its strong purgative properties, aloe sap (bitter aloes) should not be taken during pregnancy.

Medicinal and cosmetic uses

The chemical leaf constituents of *Aloe vera* and *A. ferox* are quite similar, but occur in these species in different concentrations. The key active ingredients of the leaves of these plants are the anthrone C-glucoside known as aloin (mainly found in sap), complex polysaccharides and glycoproteins (mainly found in the jelly-like pulp). The leaf sap also contains minerals, amino acids, enzymes, and other compounds in varying concentrations.

The two main leaf parts from which aloe products are obtained are the sap (or bitters) and the more or less clear, slippery inner pulp or gel. Each of these has its own uses. Leaf gel comes from fresh or dried leaf pulp. It is used as a general first-aid treatment for cuts and abrasions, minor burns, sunburn and bruises. The pulp is also used in the manufacture of health tonics, cosmetics and skin-care products, including body lotions, soaps and shampoos. Health drinks prepared from the leaf pulp are very tasty, which is quite surprising considering the extremely bitter taste of the leaf sap.

Aloe sap is obtained from the bitter, yellow leaf-juice which dries to form a dark brown resinous solid. Known in the pharmaceutical industry as 'bitter aloes' or 'Cape aloes' (*A. ferox*), and 'Barbados' or 'Curaçao aloes' (*A. vera*), the crystals can be used as is or ground into a powder. Bitter aloes is used as an ingredient in tonics and health drinks, as well as in traditional remedies taken to improve digestion or as a laxative or purgative.

Both aloe gel and bitter aloes are increasingly used in complementary medicine to treat a variety of ailments. Anthraquinones, the group of compounds comprising the 'bitter' ingredient in aloe sap, display antioxidant, anti-inflammatory and antiseptic properties, and claims have been made that they display antiviral properties.

Food and nutritional products

Aloe gel is widely used in the production of health drinks and nutritional supplements. These products normally contain about 50 per cent pulped aloe gel and are frequently flavoured or blended with mineral water or fruit juice to create a healthy, refreshing drink.

In contrast, the tonics made from bitter aloes are not always palatable and are intended to be taken in small quantities, or mixed with fruit juice to improve the taste. The unpalatable, often nauseatingly bitter, taste of several traditional remedies, such as *lewensessens*, which are known by generations of South African children, is due to the presence of aloe sap in many of these remedies.

Various culinary uses have also been found for aloe products, in the form of marmalades, jams and pickles, and it is likely that these will become more widespread as production moves from farm kitchens and home industries to larger commercial operations. Preserves (jams) made from the leaves of *A. ferox* are particularly tasty and reminiscent of those prepared from the rind of watermelons. However, it requires some skill to prepare these tasty treats.

Aloe preserve

The recipe is similar to that for watermelon preserve. Use only large, sturdy, mature leaves from *Aloe ferox*.

INGREDIENTS
For every 1 kg of aloe pieces:
1.2 kg sugar
750 ml water
8.5 ml lemon juice
A few pieces of bruised, dried ginger
Pinch of salt
Slaked lime

METHOD
1. Using a sharp knife, carefully remove the green peel from the leaf. Do this in a washing basin under a stream of running water. The extremely bitter-tasting yellow sap of aloes is contained in the peel and it is essential to remove it completely and wash away all traces of sap. Only retain the inner, opaque, jelly-like pulp.
2. Cut the pulp into chunky squares or oblong pieces. Soak the pieces in cold water overnight during which they will become somewhat glassy. Taste to make sure all the bitterness has been removed. Prick each piece several times with a fork or toothpick. Weigh the drained pieces.
3. Prepare a solution of 12.5 ml slaked lime in 3 litres of water. Let the mixture stand for a while to settle, then use the clear upper lime water to steep the aloe pieces for 24 hours. Place a plank or plate over the pieces to ensure they stay below the water. Remove aloe pieces from the lime water, rinse in cold, clean water and drain thoroughly.
4. Make the syrup, using the quantities given above for every kilogram of aloe pieces. Mix sugar, water, lemon juice, salt and ginger in a large saucepan, and bring to the boil. (Tie the ginger in a muslin bag to facilitate later removal.)
5. Place the aloe pieces one at a time into the boiling syrup. It is important to add the pieces gradually, otherwise the syrup may stop boiling. Boil the pieces for 60–90 minutes until the syrup is thick and the pieces are well cooked. Allow the pieces to cool overnight in the syrup so they can absorb as much syrup as possible.
6. The next morning, remove the aloe pieces from the cold syrup and drain them. Bring the syrup to the boil again and remove the ginger. Return the pieces to the boiling syrup, then pour into dry, sterilized glass jars, making sure no air bubbles are trapped in the mixture, and seal the jars.

Barrier plants and soil conservation

With their, sturdy, spiny leaves, aloes make excellent natural hedges and boundary fences, as well as live animal enclosures (*kraals*), and are widely used for this purpose in rural areas where they grow in abundance.

Their dense, spreading, mat-like root systems make aloes, especially those which proliferate freely by suckers, ideal for binding soil and they are used for soil stabilization, such as on mine tailings and road cuttings, and to prevent soil erosion and the formation of gulleys (*dongas*). *Aloe davyana* is often used for this purpose.

This old cultivated field on the KwaZulu-Natal side of the Mtamvuna River Valley is fenced with *Aloe ferox* plants.

In several parts of its natural distribution range, *Aloe marlothii* is grown as a fence for animal enclosures. Here, near the citrus farm 'Zebediela' in Limpopo Province, it is effectively combined with branches cut from thorn trees. The species is often used for this purpose in mixed plantings with the pencil-branched *Euphorbia tirucalli*, the milky latex of which is caustic.

Leaves from these plants of *Aloe ferox* growing wild at Aloes, an industrial development area near Port Elizabeth, were recently harvested for the extraction of their leaf exudates (see page 127).

Aloe davyana roots are used to good effect to bind the soils in an erosion-prone watercourse near Potchefstroom in North West Province. This plant has ended up on a pedestal created by the soil around its root ball being washed away. The creeping grass on the right is *Cynodon dactylon*, the indigenous couch grass (*kweekgras*) which is also widely used as a soil binder.

REFERENCES

The following is a selection of references in popular and scientific books and journals.

PART ONE: ALOES AND THEIR KIN

Introducing Aloes

MCAULIFFE, J. 2006. Ancient creosote bush clones: a trail of multidisciplinary discoveries. *Sonoran Quarterly* 60 (4): 4–8.

SMITH, G. 2003. *Eerste Veldgids tot Aalwyne van Suider-Afrika*. Struik Uitgewers, Kaapstad.

SMITH, G. 2003. *First Field Guide to Aloes of Southern Africa*. Struik Publishers, Cape Town.

SMITH, G.F. & VAN WYK, B-E. 1991. Generic relationships in the Alooideae (Asphodelaceae). *Taxon* 40: 557–581.

SMITH, G.F. 2005. *Gardening with Succulents. Horticultural Gifts from extreme Environments and the Arid World*. Struik Publishers, Cape Town.

SMITH, G.F. 2006. *Cacti and Succulents*. New Holland Publishers Ltd., London.

SMITH, G.F., VAN JAARSVELD, E.J., ARNOLD, T.H., STEFFENS, F.E., DIXON, R.D. & RETIEF, J.A. (eds) 1997. *List of southern African Succulent Plants*. Umdaus Press, Pretoria.

STEENKAMP, Y., KELLERMAN, M.J.S. & VAN WYK, A.E. 2001. Fire, frost, waterlogged soil or something else: what selected for the geoxylic suffrutex growth form in Africa? *PlantLife* 25: 4–6.

VAN JAARSVELD, E.J., VAN WYK, B-E. & SMITH, G.F. 2000. *Succulents of South Africa. A Guide to the Regional Diversity*. Tafelberg Publishers, Cape Town.

VAN WYK, A.E. & SMITH, G.F. 2001. *Regions of Floristic Endemism in southern Africa. A Review with Emphasis on Succulents*. Umdaus Press, Hatfield, Pretoria.

VAN WYK, B-E. & SMITH, G. 2004. *Guide to the Aloes of South Africa*. 2nd expanded edition. Briza Publications, Pretoria.

VOGEL, J.C. 1974. The life span of the kokerboom. *Aloe* 12: 66, 68.

WHITE, F. 1977. The underground forests of Africa: a preliminary review. *Gardens' Bulletin, Singapore* 29: 57–71.

The family Aloaceae and its genera

BRIGGS, D. & WALTERS, S.M. 1997. *Plant Variation and Evolution*, 3rd ed. Cambridge University Press, Cambridge.

HALLÉ, F., OLDEMAN, R.A.A. & TOMLINSON, P.B. 1978. *Tropical Trees and Forests: an Architectural Analysis*. Springer-Verlag, Berlin.

JEFFREY, C. 1982. *An Introduction to Plant Taxonomy*, 2nd ed. Cambridge University Press, Cambridge.

LARCHER, W. 2003. *Physiological Plant Ecology: Ecophysiology and Stress Physiology of Functional Groups*, 4th ed. Springer-Verlag, Berlin.

LAWRENCE, A. & HAWTHORNE, W. 2006. *Plant Identification: creating user-friendly Field Guides for Biodiversity Management*. Earthscan, London.

MAYR, E. 1997. *This is Biology: the Science of the Living World*. Belknapp Press of Harvard University Press, Cambridge, Massachusetts.

PANKHURST, R.J. 1978. *Biological Identification: the Principles and Practice of Identification Methods in Biology*. Edward Arnold, London.

PARACER, S. & AHMADJIAN, V. 2000. *Symbiosis: an Introduction to Biological Associations*, 2nd ed. Oxford University Press, New York.

PROCTOR, M., YEO, P. & LACK, A. 1996. *The Natural History of Pollination*. Timber Press, Portland, Oregon.

SIMPSON, M.G. 2006. *Plant Systematics*. Elsevier Academic Press, Amsterdam.

WASER, N.M. & OLLERTON, J. (eds) 2006. *Plant-Pollinator Interactions: from Specialization to Generalization*. University of Chicago Press, Chicago.

Understanding aloes

BELL, A.D. 1991. *Plant Form: an Illustrated Guide to Flowering Plant Morphology*. Oxford University Press, Oxford.

CHALKER-SCOTT, L. 1999. Environmental significance of anthocyanins in plant stress responses. *Photochemistry and Photobiology* 70: 1–9.

CUTLER, D.F. 1985. Haworthia spp. (Liliaceae) with window leaves: SEM studies of leaf surface adaptations. *Parodiana* 3: 203–223.

ENRIGHT, J.T. 1982. Sleep movements of leaves: in defence of Darwin's interpretation. *Oecologia* 54: 253–259.

GOULD, K.S., McKELVIE, J. & MARKAM, K.R. 2002. Do anthocyanins function as antioxidants in leaves? Imaging of H_2O_2 in red and green leaves after mechanical injury. *Plant, Cell and Environment* 25: 1261–1269.

JORDAN, P.W. & NOBEL, P.S. 1979. Infrequent establishment of seedlings of *Agave deserti* (Agavaceae) in the northwestern Sonora desert. *American Journal of Botany* 66: 1079–1084.

LEV-YADUN, S. 2001. Aposematic (warning) coloration associated with thorns in

higher plants. *Journal of Theoretical Biology* 210: 385–388.
NIEMELÄ, P. & TUOMI, J. 1987. Does the leaf morphology of some plants mimic caterpillar damage? *Oikos* 50: 256–257.
PAIN, S. 2002. 'Red alert'. *New Scientist* 175 (2362): 40–44.
RAUH, W. Window-leaved succulents. *Cactus & Succulent Journal* (U.S.) 46: 12–25.
SMITH, G.F., STEYN, E.M.A. & COETZEE, J. 1999. Morpho-anatomical leaf features of *Aloe suzannae* Decary (Asphoderlaceae). In: Timberlake, J. & Kativu, S. (eds) *African Plant Biodiversity, Taxonomy and Uses*, pp. 409–421. Royal Botanic Gardens, Kew.
TURNER, R.M., ALCORN, S.M., OLIN, G. & BOOTH, J. 1966. The influence of shade, soil, and water on saguaro seedling establishment. *Botanical Gazette* 127: 95–102.
VALIENTE-BANUET, A. & EZCURRA, E. 1991. Shade as a cause of the association between the cactus *Neobuxbaumia tetetzo* and the nurse plant *Mimosa luisana* in the Tehuacán valley, Mexico. *Journal of Ecology* 79: 961–970.
WEISSER, P., WEISSER, J., SCHREIER, K. & ROBRES, L. 1975. Discovery of a subterranean species of *Neochilenia* (=*Chileorebutia, Thelocephala*) in the Atacama Desert, Chile and notes about its habitat. *Excelsa* 5: 97–99, 104.
WITHGOTT, J. 2000. Botanical nursing. *Bioscience* 50: 479–388.

Aloes and extreme environments

BALUSKA, F., MANCUSO, S. & VOLKMAN, D. (EDS). 2006. *Communication in Plants: Neuronal Aspects of Plant Life*. Springer, Berlin.
BOND, W. 1983. Dead leaves and fire survival in southern African tree aloes. *Oecologia* 58: 110–114.
COGHLAN, A. 1998. Sensitive flower. *New Scientist* 159 (2153): 24–28.
COWLING, R. & PIERCE, S. 1999. *Namaqualand: a Succulent Desert*. Fernwood Press, Vlaeberg.
CRAIB, C. 2005. *Grass Aloes in the South Africa Veld*. Umdaus Press, Hatfield, Pretoria.
GIBILISCO, S. 2006. *Meteorology Demystified*. McGraw-Hill, New York.
HARDEN, P. 2003. *A Gardener's Guide to Frost*. Willow Creek Press, Minocqua, Wisconsin.
HOPKIND, W.G. & HÜNER, N.P.A. 2004. *Introduction to Plant Physiology*, 3rd ed. John Wiley & Sons, Hoboken, New Jersey.
KELLY, D. & SORK, V.L. 2002. Mast seeding in perennial plants: why, how, where? *Annual Review of Ecology and Systematics* 33: 427–447.
KOENIG, W.D. & KNOPS, J.M.H. 2005. The mystery of masting in trees. *American Scientist* 93: 340–348.
LARCHER, W. 2003. *Physiological Plant Ecology: Ecophysiology and Stress Physiology of Functional Groups*, 4th ed. Springer-Verlag, Berlin.
LINDOW, S.E. 1983. The Role of bacterial ICE nucleation in frost injury to plants. *Annual Review of Phytopathology* 21: 363–384.
PHILLIPS. H. 2002. Not just a pretty face. *New Scientist* 175 (2353): 40–43.
SATO, J. 2003. Weather change and pain: a behavioral animal study of the influences of simulated meteorological changes on chronic pain. *International Journal of Biometeorology* 74: 55–61.
SCHOLES, M.A. 1988. A population study of *Aloe peglerae* in habitat. *South African Journal of Botany* 54: 137–139.
TAIZ, L. & ZEIGER, E. 2006. *Plant Physiology*, 4th ed. Sinauer Associates, Sunderland, Massachusetts.
THOMAS, P.A. & GOODSON, P. 1986. How do succulents cope with fire? *British Cactus and Succulent Journal* 4: 111–112.
VAN ZYL, D. *South African Weather and Atmospheric Phenomena*, Briza, Pretoria.
WEISSER, P.J. & DEALL, G.B. 1989. *Aloe petricola*: ecological notes and effect of fire near Sabie. *Aloe*: 26: 27–30.
ZÜRCHER, E., CANTIANI, M-G., SORBETTI-GUERRI, F. & MICHEL, D. 1998. Tree stem diameters fluctuate with tide. *Nature* 392: 665–666.

PART THREE: GARDENING WITH ALOES

BROWN, J.R., WHITE, A., SLOANE, B.L. & REYNOLDS, G.W. 1939. *Succulents for the amateur. Introducing the beginner to more than eight hundred fascinating succulent plants*. Abbey Garden Press, Pasadena.
KOELEMAN, A. 1962. Die veredeling van die aalwyn. *Lantern* 12,2: 82–91 [English summary: p. 91].
SMITH, G.F. & CORREIA, R.I. DE S. 1988. Notes on the ecesis of *Aloe davyana* (Asphodelaceae: Alooideae) in seed-beds and under natural conditions. *South African Journal of Science* 84: 873.
SMITH, G.F. & CORREIA, R.I. DE S. 1992. Establishment of *Aloe greatheadii* var. *davyana* from seed for use in reclamation trails. *Landscape and Urban Planning* 23: 47–54.
SMITH, G.F. & VAN WYK, A.E. 1996. Arthur Koeleman (1915–1994). *Bothalia* 26: 77–80.

PART FOUR: USES OF ALOES

ROOD, B. 1994. *Kos uit die veldkombuis*. Tafelberg, Cape Town.
TATE, J.L. 1978. *Cactus Cook Book*. Cactus & Succulent Society of America, Reseda, California.

Aloe pluridens

INDEX

Page numbers in *italics* refer to photographs.

A

aalwyn 7
Adansonia digitata 7
Acacia 83
Afromontane forest 73
Agave 8, 22, *23*, 36, 37, 38, 79
Agave americana 31, *49*
Agave attenuata 113
Agave gigantensis 31
Agave xylonacantha 10, *31*
air humidity 106
Albizia 83
Aloaceae 7, 15, 17
Aloe 7, 8, 9, 10, 11, 15, 18, 57, 73, 83, 96, 105
Aloe family key 26–27
Aloe genera 17–21
Aloe characters 10–11
  cold tolerance 49–52
  fire tolerance 44–48
  intelligence 52–53
  life span 12
  'masting' 52–53
aloe parts
  anchorage 29
  flowers 38–41
  fruits 38–41
  inflorescences 38–41
  leaves 29–37
  roots 29
  seeds 42–43
Aloe aculeata 31
Aloe affinis 84
Aloe africana 43, 68, *135*
Aloe alooides 74
Aloe angelica 82, 84
Aloe arborescens 2, 6, *17*, 39, 43, *96*, 97, *102*, *104*, *109*, *110*, 112, *122*, *125*, 128
Aloe aristata 123

Aloe barberae 13, *14*, 30, 73, 74, *104*, 112, *123*
Aloe bowiea 11
Aloe branddraaiensis 85
Aloe brevifolia 37, 46, *54*, 64, 107
Aloe broomii 41
Aloe bulbicaulis 11, *39*
Aloe bulbillifera 11
Aloe burgersfortensis 85
Aloe camperi 96, 98, *109*
Aloe candelabrum 86
Aloe castanea 83, 87
Aloe chabaudii 87, 113
Aloe chortolirioides 47
Aloe ciliaris 69
Aloe commixta 46, 64, 107
Aloe compressa 31
Aloe comptonii 15, 65, *112*
Aloe cryptopoda 38, 88, 90, *125*
Aloe davyana 78, 98, 107, 109, 128, 129, *131*
Aloe dichotoma 12, *13*, 14, 30, *56*, 58, 59
Aloe dorotheae 34, 36
Aloe ecklonis 77
Aloe excelsa 88
Aloe ferox 1, *17*, 31, *42*, 43, 52, 58, *59*, 86, *102*, 127, *127*, 128, 129, *130*, *131*
Aloe fibrosa 36
Aloe forbesii 37
Aloe globuligemma 10, *30*, 89
Aloe graciliflora 78
Aloe grandidentata 99
Aloe greatheadii 78
Aloe greenii 68
Aloe haemanthifolia 30, 36, 46, 107
Aloe immaculata 89

Aloe lineata 69, *106*
Aloe littoralis 32, *125*, 134
Aloe lutescens 90
Aloe maculata 100, 109
Aloe marlothii 2, *4*, *8*, *9*, 10, 12, 31, *41*, *43*, *47*, 47, 52, 88, 90, 93, *103*, *110*, *124*, 128, *131*
Aloe micracantha 30, 46
Aloe microstigma 43, *44*
Aloe ortholopha 121
Aloe parvibracteata 41, 91
Aloe peglerae 45, 78, *123*
Aloe perfoliata 46, *54*, 65, *117*
Aloe petricola 119
Aloe petrophila 46
Aloe pictifolia 109
Aloe pillansii 48, 60
Aloe plicatilis 30, 36, 46, *47*, 66, 107
Aloe pluridens 70, *133*
Aloe polyphylla 50, 51, 79, 107
Aloe porphyrostachys 10, *41*
Aloe pratensis 80

Aloe pretoriensis 91
Aloe pruinosa 92
Aloe ramosissima 60, *61*
Aloe rupestris 72, 75
Aloe speciosa 40, *67*, 70
Aloe spectabilis 93
Aloe spicata 43, *55*, 93
Aloe spinosissima 114
Aloe 'Spiraal' *40*
Aloe squarrosa 10, *28*
Aloe striata 71, *113*, *119*
Aloe striatula 49, 80, *107*, 113
Aloe succotrina 66
Aloe suffulta 109
Aloe suprafoliata 31, 81
Aloe suzannae 11
Aloe tenuior 71, *118*
Aloe thraskii 49, 68, 75, *106*
Aloe transvaalensis 94, 107
Aloe vanbalenii 35, 36
Aloe vanrooyenii 83, 95
Aloe variegata 40, *51*, 61, 119, *124*
Aloe vera 101, *107*, 127, 128

Aloe littoralis

Aloe verecunda 46, 81
Aloe vossii 38
Aloe wickensii 90, 95
Aloe zebrina 94
annual 12
anthocyanins 36
Arabian Peninsula 9, 101
Asphodelaceae 15
Asteraceae 57
Astroloba 18
Astroloba foiliosa 19
Astroloba rubriflora 11, 19, 36
Astroloba spiralis 19

B
baobab 7
bergaalwyn 128
biennial 12
'bitter aloes' 128
Burgersfort 83, 85
bushveld (*see also* savanna) 54, 83

C
'Cape aloes' 128
Cape Floristic Region 45, 63
Cape shrublands (*see* fynbos)
'Capensis' 63
chemotypes 16
Chortolirion 19
Chortolirion angolense 19, 47
classification 15, 16
coastal thicket (*see also* thicket) 97
commensalism 25
companion plants (*see* nurse plants)
Crassula rupestris 8
Crassulaceae 63
cross-pollination 121
cultivars 16
Cussonia spicata 67
cuttings 122

D
daisy family 57
desert 54, 57
diseases 124–125
Doryanthes 11
drainage 109
Drakensberg 49

E
East Africa 98
Eastern Cape 20, 30, *30*, 45, 49, 52, 67
ecotypes 16, 107
Ericaceae 63
Euphorbia mauritanica 8

F
fire 45–48
forest 54, 73
Frithia pulchra 33
frost 49, 50–51, 52
frost tolerance 49–52, 108
Furcraea 22, *23*
fynbos 45, 46, 48, 54, *62*, 63, 97, 100
Fynbos Biome 45, 63

G
Gasteria 20, 32, 61
Gasteria acinacifolia 20, 32
Gasteria armstrongii 30
Gasteria distichia 30
Gasteria maculata 30, 32
Geraniaceae 63
germination 120–121
grass aloes 77, 81, 83, 109
grassland 54, 76, 97, 98, 99, 100
Great Escarpment 73, 76
Great Karoo 45, 49
growing aloes 105–109

H
habitat 54, *55*
Haworthia 18, 19, 21, 32, 33
Haworthia attenuata 33
Haworthia fasciata 21
Haworthia glabrata 21, 33
Haworthia maughanii 33
Haworthia maxima 33
Haworthia radula 33
Haworthia truncata 33
Hesperaloe 24
Hexandria Monogynia 15
Hesperaloe parviflora 24
Highveld 49, 76

I
ice formation 50–51
identification key 26–27
irrigation 105

Aloe africana x *Aloe ferox*

J
Joshua tree 24

K
Kalahari 57
Kalanchoe sexangularis 34
Karoo 57
karroid vegetation 99
Kniphofia spp 30
KwaZulu-Natal 20, 45, 49, 52, 67

L
Larrea tridentata 12
leaf spines 31
Lesotho 49, 50, 51, 76, 79, 80
light 108
Liliaceae 15
Linnaeus, Carl 15
Lomatophyllum 9, 11, 18, 40

M
maculate aloe 31, 84, 85, 89, 94, 95, 123
Madagascar 9, 18, 31
Mascarene Islands 9, 18
Mediterranean climate 48, 63, 65, 106
Mesembryanthemaceae 57, 63
Mexico 22, 24
mimicry 33
monocarpic 12
Mpumalanga 49
mutualism 25

N
Namaqualand 57, 58
Namib Desert 57
non-discriminating aloes 96
nurse plants 43, 120

P

parasitism 25
perennial 7, 12
pests 124–125
Poellnitzia 18, 36
polycarpic 7, 12
pot plants 115–117
Pretoria National Botanical Garden *6*
prickles 37
propagation 123
Proteaceae 63
quiver tree 12

R

rainfall 105
repotting 115
Restionaceae 63
Richtersveld 57

S

SANBI *6*
Sanseveria hyacinthoides 32
savanna 54, 57, *82*, 83, 97, 98, 99, 100
Sedum sediforme 8
seeds 43, 119–121
Sekhukhuneland 83
semi-desert 54, 57
Senecio rowleyanus 120
shade 108, 109
Skirt aloe 74
Socotra 9, *28*
soil 109, 116
soil conservation 130–131
Sonoran Desert 12
South America 22
southwestern Cape 19
Soutpansberg 84
spines 37
spotted aloe (*see also* maculate aloe) 84
subtropical east coast 97, 98, 100
subtropical forest 73
Succulent Karoo 48
succulents 7, 22
sunlight 108
Swaziland 31
symbiosis 25

T

taxonomy 16
temperature 106
thicket 54, 67, 97, 100
thorns 37
Tree aloe 74
tropical forest 73

U

USA 24
uses of aloes 126–129

V

valley bushveld (*see* thicket)
vygie family (*see also* Mesembryanthemaceae 57

W

waaier-aalwyn 46
watering 105
window-leaves 33
Witwatersrand 76
Worcester-Robertson Karoo 19, 30

X

Xanthorrhoeaceae 15

Y

Yucca 24
Yucca brevifolia 24
Yucca desmetiana 24
Yucca elephantipes 25

Z

Zimbabwe 19

The G.W. Reynolds Gate at the Pretoria National Botanical Garden depicts a number of aloes in wrought iron. The gate was named for Dr Gilbert Westacott Reynolds, who contributed substantially to our understanding of these plants.